# The Shattered Self

# The Shattered Self

## The End of Natural Evolution

Pierre Baldi

A Bradford Book
The MIT Press
Cambridge, Massachusetts
London, England

This book was set in Bembo by Achorn Graphic Services, Inc. and was printed and bound in the United States of America.

Library of Congress Cataloging-in-Publication Data

Baldi, Pierre.
The shattered self : the end of natural evolution / Pierre Baldi.
p. cm.
"A Bradford book."
Includes bibliographical references and index.
ISBN 0-262-02502-7 (hc.: alk. paper)
1. Genetic engineering. 2. Genetic engineering—Moral and ethical aspects. 3. Human Genome project. 4. Human Genome project—Moral and ethical aspects. I. Title.
QH422.B355 2001
599.93—dc21

00-048183

To B and MJ

# Contents

# Preface

This book is not meant primarily for scientists, but for laymen interested in who we are and where we might be headed. It was conceived in primitive form several years ago and perhaps would have never been published without the striking progress recently achieved in computer and biological sciences and technologies. A week does not go by these days without a major announcement.

Most of the technical background and other information that is not essential for understanding the main points of the book has been placed in a number of notes and appendices. These can easily be skipped in a first reading.

While writing the book, I have received help and support from many family members, friends, and colleagues. Pierre-François Baisnée, Søren Brunak, Paolo Frasconi, Anton Schmitt, and especially Brigitte Baldi provided me with very detailed feedback and numerous useful suggestions. The term "fiction science" was coined by Søren. Additional family members, friends, students, and colleagues I would like to thank for helping me in one way or another include Lorraine Baldi, Mario Blaum, Mark Borodovsky, Yves Chauvin, Giorgio Dalla Volta, Maria Gabriella Di Benedetto, Olivier Doria, Sachiko Eguchi, Loren Elliott, Daniela Flohr, Juan Gotrell, Donna Lew, Ana MacMassey, Van Mittal-Henkle, Ascanio and Tanja Pignatelli, Gianluca Pollastri, Burkhard Rost, Gill Williamson, and Tess Sarayba. I am grateful to my cousin Gilles Larrain and David Teplica for the use of their photographs and to the entire staff at MIT

Press and in particular to my editors, Amy Brand and Robert Prior, and to Katherine Almeida, Sara Meirowitz, and Ann Rae Jonas. Sarah Jeffries provided editorial support and Phalem Hurewitz assisted me on contractual matters. I would like also to thank several institutions where portions of this book were conceived or written: the California Institute of Technology, the Center for Biological Sequence Analysis of the Technical University of Denmark, the Newton Institute in Cambridge, England, the University of Florence, Net-ID, Inc., and the University of California, Irvine, where the book was completed. I am particularly grateful to UCI and to my new friends and colleagues for providing a supportive and stimulating environment.

*We do not know who we are, but we know enough to know we are not who we think we are.*

# Introduction

# 1

*Learning to thrive in constant change is the next frontier. . . . There is no place to hide from change. Nor should you want to, as change promotes growth.*

—Michael Dell, founder and CEO, Dell Corporation

At the beginning of the third millennium, mankind for the first time stared at its genome, the blueprint out of which human flesh is made. The most significant event in our history, and perhaps in the 3.5 billion-year history of life on Earth, occurred during a time of general euphoria driven by computers, the Internet revolution, and one of the longest periods of sustained economic expansion. Less than 50 years had passed since Watson and Crick's discovery of DNA's fundamental structure and how genetic information is encoded.

Today, the Human Genome Project, the largest and boldest scientific project ever undertaken, has produced nothing less than the complete sequence of genetic instructions, the program that orchestrates who we are at the molecular level. Each and every one of us started the journey through life as a single cell. It is this program that progressively coordinates the division, specialization, and assembly of trillions of cells into each one of us. It is this program, in one of its many possible variations, that is ticking in each one of our cells, directing all its activities, from our conception until our last breath.

The completion of the Human Genome Project opens a whole new set of possibilities for biology, biotechnology, medicine, and society at large. A similar revolution occurred when humans first learned to read and write.

We are now learning to read and write the language we are made of. But in many ways, it is just a beginning and only a piece of a larger revolution that is well under way. Within the same generation, the human brain will face machines that surpass its raw computing power and an interconnected world of information-processing devices that makes science fiction pale in comparison. Together these milestones raise profound and troubling questions about the nature and boundaries of life, intelligence, and who we really are. These questions are the fabric of this book.

It may comfort some to realize that modern biotechnology, which plays a central role in the book, is in many ways just a powerful extension of human agricultural and breeding practices. Such practices have been going on for at least many thousands of years. Throughout our history, we have used breeding to manipulate and select the genomes of plants and animals, and even of ourselves. After all, at the beginning of 1999 interracial marriages were still against the law in the state of Alabama. For most of our past, such control could be exerted only at the macroscopic level of entire organisms. It was crude, slow, and cumbersome. Now we begin to be able to manipulate genomes directly at the microscopic level, the level of single genes and their constituents, with unprecedented speed and precision. This ability is greatly accelerating the pace of change and discoveries as well as enlarging the realm of possibilities; hence it brings with it a sense of revolution.

To explore these changes, we need only extrapolate a little from current technology and look to several examples slightly ahead in the future. No science fiction here—these examples do not rely on any new fundamental principles or inconceivable amounts of ingenuity, and certainly on nothing like aliens. What I am talking about is "fiction science," extensions of existing technology. They can be viewed as scientific with a high degree of plausibility, and most scientists would agree that, with sufficient funding and research, such technological feats are feasible. The general path on which to reach them is already somewhat clear.

Thus the book looks at what is scientifically possible. But the possibilities alone have far-reaching consequences. Their contemplation challenges our most cherished but naive beliefs. Consider, for instance, cloning. It has already been done with mice, sheep, cows, pigs, and monkeys. Undoubtedly it can be extended to humans. In fact, from a purely technical standpoint, it could probably be achieved in a few months. In a real sense, it is just around the corner. Yet human cloning challenges the stability of who we really are, regardless of whether or not we end up doing it.

And the truth is that we are not well prepared for human cloning. Many of our discussions avoid the topic, or center on technological and political aspects. For all its fascination, cloning seems to make our minds go blank. It makes us feel very uncomfortable somewhere. This uneasiness comes from the semiconscious realization that it stands the chance of challenging the idea of who we are. But cloning is here, and here to stay; it is pointless to try to avoid it. We might as well make the effort to stir our minds in its direction and broaden the debate if we want to understand its meaning and consequences.

In fact, our notions of self, life and death, intelligence, and sexuality are very primitive and on the verge of being profoundly altered on the scale of human history. It is this shattering—what causes it and its meaning—that forms the central thread of this book. This shattering is brought about by scientific progress in biology, computer science, and resulting technologies such as biotechnology and bioinformatics. The first cracks and fissures have appeared all around us. They are progressively infiltrating our lives and societies in many ways; however, it may take decades, if not centuries, of further technological refinements for the shattering to be complete and fully realized.

Through millions of years of evolution our brains have been wired to provide us with an inner feeling of self, a feeling that each of us is a unique individual delimited by precise boundaries. We view each person as endowed with special feelings and thoughts, and fundamentally different from other beings, other animals, or, for that matter, other information-processing systems such as computers. We have also been wired to reproduce ourselves in a certain way. These concepts are central to ourselves and to the way we function, and are reinforced by our education and culture.

A fundamental argument of this book is that this self-centered view of the world is problematic—in fact, it is "scientifically" wrong. It is the result of evolutionary accidents. The reason for its past success lies in being an adequate model of the world during our evolutionary bootstrapping: a world without molecular biotechnology, human cloning, and the Internet. A world where identical twins and Einsteins are considered rare and unusual. A world with no computers. A world in which sexes and species are well separated, and sexual activity is necessary to reproduce and mix genomes. We have lived beside the tip of an iceberg and aligned our views with the corresponding environment. The reality of the world has remained hidden under water up to now, but today the bulk of the iceberg is starting to emerge. In fact, it is beginning to tower over us.

In time, we are going to have to come to terms with the fact that we have had the wrong model, or rather, the wrong data. As we shall see, genomes, computations, and minds are rather fluid and continuous entities, both in space and in time. Individually, we are just samples of this continuum. Myriad other selves are arbitrarily close to ours, selves continuously interpolated between ourselves and any other being, including those of the opposite sex. Both brains and genomes can be morphed, and the world is home to other creatures and processing systems, biological and other, most of which we have never seen before. We can raise our clones and have multiple brains. The boundary between the self and the other, the self and the world, the inside and the outside has begun to blur, and ultimately may evaporate entirely.

The only way to appreciate what could happen is to understand what is happening now and project it into the future. Yet the essence of this book is *not* about predictions. Although some of the fiction science examples are likely to occur, what matters most for the present perspective is that such events are *possible.* The closest analogy I can think of is nuclear war. The knowledge that it is technically feasible to build nuclear weapons profoundly changes the way we see the world. This would remain true even if we had decided not to build these weapons in the first place. The possibility of an all-out nuclear holocaust has changed us forever even if, so far, we have been wise enough to avoid it.

The book includes a summary of the basic scientific background required to understand the realm of possibilities within our grasp, and their implications. As with every amazing time of scientific progress, we ought to be aware of the consequences brought about by the new technologies, and the ethical issues they inevitably raise. These must be faced not in fear, but with knowledge and reason. The central thread of the book, however, is not about ethics, or politics, or what should and should not be done. It is above all about who we are, regardless of our values, past, present, or future. To follow this thread, the reader should put aside feelings and ethical values for a time, because before passing a judgment or deciding on a course of action, it is necessary to have a clear view of all the underlying facts and possibilities. Only if we strive to examine objectively where we might be headed, do we stand a chance of making rational decisions. History has shown us many times that aligning our models of the world with reality in the end leads to more beneficial solutions.

It cannot be denied, however, that the topics covered in the book may be perceived as disturbing, unsettling, and threatening—perhaps even shocking. This is no wonder: challenging our identities and boundaries generates profound uneasiness, or even repulsion, and comes at a price. It is the same feeling we had to face at each stage of scientific progress; for instance, when we had to realize that our planet was neither flat nor at the center of the universe. The uneasiness that comes with the unknown and with having to change our view of the world. Life does not really feel the same once we begin to see ourselves as "biological information-processing machines." In many ways, our genetic makeup and education go against such ideas. I do not advocate any of the fiction science scenarios, which I find as disturbing as anyone else does. But exploring the universe, regardless of whether or not it is likable, is also part of the human quest. The things we do not want to look at are sometimes the most revealing of who we are. They will ultimately shatter the last illusions that have accompanied us during our evolutionary bootstrapping.

Figure 2.1 A seventeenth-century depiction of sperm, made at a time halfway through the decentering process: scientific enough to use a microscope, avoid religious overtones, and propose an empirical approach, but fundamentally wrong and self-centered.

# Science Today

# 2

*Science is the most exciting and sustained enterprise of discovery in the history of our species. It is the great adventure of our time. In a stunningly short period of time, science has extended our knowledge all the way from the behavior of galaxies to the behavior of particles in the subatomic world.*

—Michael Crichton, writer[1]

*Religion is something left over from the infancy of our intelligence; it will fade away as we adopt reason and science as our guidelines.*

—Bertrand Russell, philosopher and mathematician

Before we look at what may be likely to happen, it is useful to focus on the state of science today, where we are and how we got there. This provides us with a sense of where the boundaries of our knowledge are located. It is precisely at those boundaries that science and then fiction science begin.

Many scientists would agree that most of the fundamental principles in science are well understood, and that many of the remaining important open questions are concentrated in biology and computer science. There is little doubt that biology and computer science will be two dominating sciences of the first few decades of the twenty-first century.

Indeed, for centuries, mathematics provided the language and foundation for all other exact sciences, especially physics, and with great success. Today, the basic mathematical principles behind the majority of our technological applications are thoroughly understood. Though mathematics is by its essence open-ended—new abstract axioms and theories can always be created—many, if not most, of its classical conjectures have been solved. Even Fermat's famous last theorem was proved by Andrew Wiles in 1994.[2]

One of the most important open problems in mathematics is in fact a computer science problem. It is the so-called P = NP problem. This is essentially the question of whether classes of problems exist that are solvable in principle,[3] but for which an exact solution cannot be found efficiently on a digital computer. The classic candidate problem in this class is the so-called Traveling Salesman Problem. Given N cities and their pairwise distances, find the shortest tour visiting all cities only once. This is a finite problem with an obvious solution: list all possible tours, measure the total length of each tour, and then select the shortest one. The difficulty lies in the combinatorial explosion—the total number of possible tours grows exponentially with the number N of cities.[4] Listing all possible tours of a few hundred cities is beyond the reach of any existing computer. The P = NP conjecture corresponds to the widespread belief that an efficient[5] algorithm for the exact solution of such problems does not exist. Little progress has been made toward resolving this conjecture since it was first introduced in the early 1970s.

Supported by a strong mathematical foundation, physics and chemistry have made enormous progress over the last four centuries. The remarkable technological advances they have spearheaded are another strong indication that all the fundamental principles are in place. Physics has been capable of accounting for most phenomena, from the subatomic scale with quantum mechanics to the cosmic scale with general relativity, all with a high degree of precision. This is not to say that there are no more problems to be solved in physics or chemistry, but that the majority of them have little to do with fundamental principles.[6] Most unsolved problems have to do with two types of situations. First is the case of extreme conditions, strange states of matter, to which we have limited experimental access, such as very high or low temperatures, pressures, or magnetic fields. This is also the case with the first few moments right after the Big Bang that originated our current universe. The second class of open problems has to do with the emergent properties of large complex systems, as in statistical mechanics and solid-state physics. When systems become large and have many degrees of freedom, as in the Traveling Salesman Problem, predicting the macroscopic system properties from the well-understood microscopic principles becomes difficult or not always computationally feasible. In fact, the reduction of chemistry to physics falls largely in this class. Although theoretically on clear ground, the precise reduction of complex chemical reactions to quantum mechanics remains intractable on current computers.

Finally, living systems are exceedingly complex, in part as the result of billions of years of evolutionary tinkering. We do not completely understand all their system-level properties—for instance, how each gene is regulated, how embryos develop, how immune systems function, and how the human brain processes information. These are precisely examples of large systems with many interacting elements, and with macroscopic emergent and self-regulating properties. But biophysics and biochemistry paved the way for molecular biology, and it is fair to say that today we understand the basic principles of life at the molecular level. No new special concepts, beyond those of physics and chemistry, are necessary to explain the phenomena of living matter.[7] Furthermore, as we are about to see, molecular biology has given us a firm foundation on which to build so we can progress toward the understanding and manipulation of biological systems.

## The Decentering Process

*I have in fact imagined, that I could say as I beheld the animalcules in the semen of an animal that there lies the head and there as well the shoulders and there the hips; but since these notions have not the slightest shred of certainty, I will not yet put forward such a claim, but hope that we may have the good fortune to find an animal whose male seed will be so large that we will recognize within it the figure of the creature from which it came.*

—Antoni van Leeuwenhoek, janitor, microscopist, and scholar (1685)

*Could not the device that opens the Eustachian canal of the deaf, open that of apes? Might not a happy desire to imitate the master's pronunciation liberate the organs of speech in animals that imitate so many other signs with such skill and intelligence? Not only do I defy anyone to name any really conclusive experiment that proves my view impossible and absurd; but such is the likeness of the structure and functions of the ape to ours that I have very little doubt that if this animal were properly trained, he might at last be taught to pronounce, and consequently to know a language. Then he would no longer be a wild man, now a defective man, but he would be a perfect man, a little gentleman, with as much matter or muscle as we have for thinking and profiting by his education. . . . The transition from animals to man is not violent, as true philosophers will admit. What was man before the invention of words and the knowledge of language? An animal of his own species with much less instinct than the others.*

—Julien Offray de La Mettrie, physician, *Man a Machine* (1747–48)

It is also useful to have a sense of the major historical trends behind the expansion of our knowledge, because these trends are likely to continue in the future. What is clear is that the history of Western science and civilization has been a history of progressive *decentering,* a gradual movement away from a self-centered view of the world that comes so naturally to us.

In a nutshell, our distant ancestors first thought that Earth was flat and contained the entire universe. After realizing the existence of other planets and stars, they found refuge in the idea of Earth at the center of the universe. Then, as late as the Middle Ages, Earth was commonly considered to reside at the center of the solar system. Only five centuries ago, during the Renaissance, did modern science begin to flourish together with the use of experimental methods. This idea that one could learn something about the nature of the universe by carrying out experiments was by itself a breakthrough. Before that, the truth had to be found by theoretical considerations only, or by authority—the study of the Bible, the work of ancient philosophers, and so forth.[8] With great resistance, it was progressively realized and finally accepted that our planet is not flat but roughly spherical, and that it does not lie at the center of anything, especially not at the center of our planetary system. The sun lies at the center of the solar system, and Earth is one of nine major planets orbiting the sun. The solar system itself is only a small system within the Milky Way, our galaxy, which contains over 100 billion other stars. In fact, the solar system is more toward the periphery of the Milky Way than toward its center. In turn, there are over 100 million galaxies in the universe as a whole. This decentering process toward the cosmic scale of the universe had its counterpart in the decentering process toward the microscopic scale.[9]

A similar process occurred for other dimensions, such as time. Time was initially thought of as being an absolute quantity, for instance, independent of speed, and more or less limited to the scale of human generations and human history. We have progressively become aware of geological, evolutionary, and astronomical time scales measured in billions of years, well beyond the tiny 1-million-year horizon of human evolution. At the other end of the spectrum, we have discovered that neuronal electrical activity evolves in milliseconds. We have built computer clocks that run in billionths of seconds. We have observed chemical reactions in the making, using femtoseconds-long (1-millionth of a billionth of a second) laser pulses. Einstein's theory of relativity showed beyond any reasonable doubt the intrinsically relative nature of time.

Decentering also played a central role in biology, perhaps even more so, because biology has to do with what we are made of. We have gone through many phases, starting from the most centric, vitalistic, and magical perspectives of the biological world to our current views. In a vitalistic perspective, for instance, life is a special phenomenon that obeys its own laws, which are different from those obeyed by matter. It results from a special form of energy—vital energy. In ancient times we thought we were made of a stuff completely different from, say, rocks, or that we were very different from other animals or plants, and that we occupied a special position in nature. Our ancestors came up with many strange mechanisms to explain physiological processes such as disease and reproduction. A diseased man was "possessed by demons." The first key invention toward a scientific understanding of life was probably the microscope. Only after that were scholars in the seventeenth century, such as Antoni van Leeuwenhoek, able to observe bacteria, blood cells, and sperm, and build the first—sometimes baroque—theories of how these "animalcules" could function (see figure 2.1). Still, as late as the first part of the twentieth century, eminent physicists such as Erwin Schrödinger suggested that living organisms obeyed special laws and were endowed with "vital" energy.

Today, there is little possible doubt regarding these matters. The basic tenets of molecular biology and Darwin's theory of evolution have been proved true beyond any reasonable doubt. And this by many lines of evidence, from the fossil record to DNA sequence comparison. The most basic aspects of life are understood at the molecular level and do not require the introduction of new laws with respect to physics and chemistry. The same laws govern the animate and the inanimate. *Homo sapiens* is made of atoms, like all other macroscopic objects we know, and at the molecular level is similar and related to all other living organisms. We share DNA, the basic mechanism for encoding and transmitting genetic information, with crocodiles, bugs, flowers, and trees.

Each decentering step led to a higher level of abstraction and a better understanding of the universe. Historically each step, however, encountered a great deal of resistance and was associated with a (Copernician) revolution, a relatively abrupt and substantial revision in the ways humans think and see the world. Yet remarkably, in spite of all the scientific breakthroughs and technological progress, for many areas of life we have retained a centric view of the world. At the psychological level, in particular, we view ourselves, each one of us, as a special and unique entity endowed with special feelings, intelligence, and thoughts, connected to but also separated from other beings or other information-processing entities. In other words, in our daily lives, we still often view our "self" as the center of the flow of information we process. But the last wave of decentering has just begun. And, as we will see, it will ultimately shatter these last illusions that have accompanied us during our evolutionary journey.

## The Psychology of Large Numbers

Large numbers play an important role in the decentering process and throughout this book. There is something special about large numbers and our ability or inability to grasp them. In fact, the range of large numbers commonly used in a civilization is strongly correlated with that civilization's degree of technological advancement. We have come a long way from the invention of the first digits, associated with the fingers of our hands. The largest numbers commonly used in a concrete way in our civilization are in the range of billions to tens of trillions ($10^9$ to $10^{13}$). The prefix "tera," which is sometimes used, as in terabytes, corresponds to 10.$^{12}$ The prefix "peta" ($10^{15}$) is not common, but is starting to appear in technical literature. Naturally, scientists do manipulate larger numbers,[10] but in a way that is more abstract and specialized. The $10^9$ to $10^{13}$ range is the size in bytes of computer memories (1 gigabyte is 1 billion bytes) and the size in dollars of the largest economies today. At the time of this writing, Bill Gates's fortune, corresponding to less than one-third ownership of Microsoft, oscillates in the range of 50 to 100 billion dollars ($10^{11}$). The United States economy is about 10 trillion ($10^{13}$) dollars. In fact, how tangible a given range is, depends on what is being measured. It is relatively easy to form a mental picture of a billion dollars. A billion years—the scale of slow evolutionary events—on the other hand, is a much more difficult concept to grasp. When it comes to humans with identical genomes, we are back to a very primitive state. The largest number most of us are familiar with is two, the case of identical twins, and even this is a relatively rare event.

In this book, we must make an effort to look at numbers beyond the ranges we are comfortable with, such as computer memories in the range of $10^{13}$ to $10^{20}$, or 1 million clones of the same individual. We have to explore the total information contained in the human genome as well as the finite, albeit large, capacity of our brains. Indeed, our sensory systems, our attention, our memory have finite horizons that for a long time could not be concretely perceived and sized. The number of memories a human brain can hold, or the total number of inputs received by a brain over a lifetime, is a very large number we are not accustomed to think of. Furthermore, such numbers are beyond the realm of the way we normally operate psychologically. Most of our brain processing occurs outside our awareness; and, for instance, when we access our memory, we do not usually worry about how much space is left, or what can be deleted from its "directories." As we will see, all this is on the verge of changing as information-processing systems are closing in rapidly. Simple upper boundaries for such large numbers associated with our functions and our brains exist, and are getting within the reach of silicon-based computing devices.

To get to these large numbers most of us are not accustomed to, we will use over and over a simple technique informally called "back-of-the-envelope calculation."[11] For instance, we will estimate the total number of human beings who have ever lived, or the number of bits representing all the inputs received by a human brain over an entire lifetime. The approach typically simplifies the problem and concentrates only on orders of magnitude. As an example, consider the total number of neurons in the human brain, which is very large but finite. The precise number is not known, but is believed to be somewhere between 1 trillion and 1,000 trillion ($10^{12}$ to $10^{15}$). It is easy to see with a back-of-the-envelope calculation that $10^{15}$ must be a rough upper bound. Neurons are cells with diameters that are typically in the range of several microns (a micron is 1 millionth of a meter). Therefore a trivial lower bound on the volume of a neuron is a cubic micron. A rough approximation to the volume of the brain is a cube of 10 centimeters on each side.[12] This immediately yields a brain volume of 1 liter, or $10^{15}$ cubic microns, so the brain cannot have more than $10^{15}$ neurons at any given time. This upper boundary can easily be refined to a better estimate in the range of 1 trillion neurons used in the rest of the book.[13]

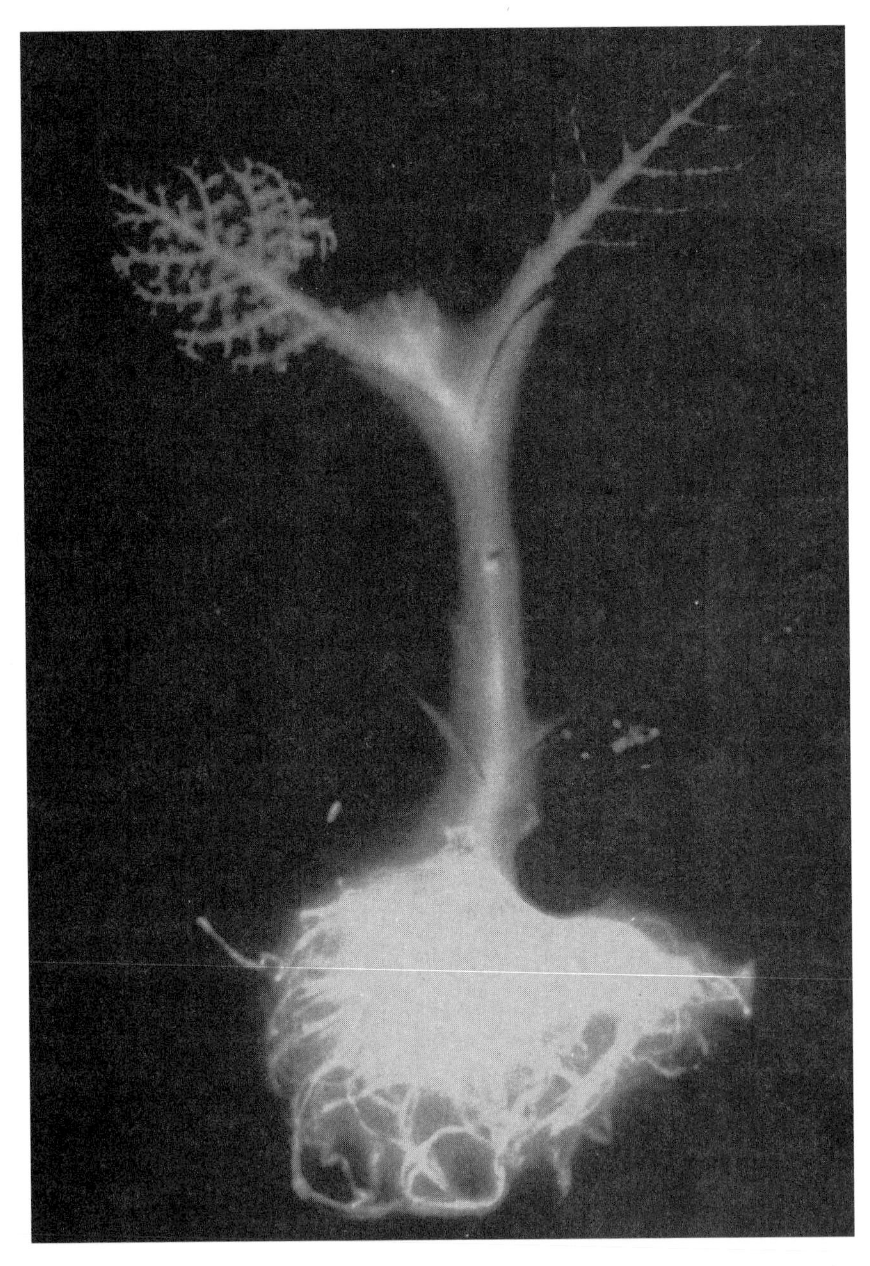

Figure 3.1 A plant that glows in the dark: gene transfer from animal to plant. A firefly gene coding for the enzyme luciferase has been inserted into this tobacco plant. The enzyme breaks down the chemical luciferin—added to the water given to this plant—releasing light in the process. (Courtesy of Visuals Unlimited.)

# Molecular Biology

# 3

*All living things have much in common, in their chemical composition, their germinal vesicles, their cellular structure, and their laws of growth reproduction. Therefore I should infer that probably all the organic beings which have ever lived on the earth have descended from some one primordial form.*

—Charles Darwin, *On the Origin of Species* (1859)

Throughout the twentieth century, especially its second half, molecular biology underwent remarkable expansion. It clarified the fundamental principles of life at the level of molecules. It laid a solid foundation for many other branches of biology and biotechnology, and gave us the means to understand and manipulate DNA, the carrier of genetic information, the blueprint of biological organisms.

The end of the twentieth century saw a remarkable acceleration in genome-sequencing projects, across the kingdom of life, that continues today. In addition to hundreds of viral genomes, dozens of bacterial genomes have been sequenced. The first genome of a complex cell, yeast, was completed in 1996, and that of the first multicellular organism, a small roundworm, in December 1998. Most remarkably, the complete genomes of complex organisms such as the fruitfly and man became available at the beginning of the new millennium. But to fully understand the implications of these breakthroughs, it is necessary to understand the basic principles of life at the molecular level.

At the molecular level, life relies very substantially on two main types of molecules: DNA with its related RNA, and proteins.[1] To a first approximation, both DNA and proteins can be viewed as long chains or sequences of basic repetitive building blocks. Both DNA and proteins fold in complex three-dimensional structures. It is the particular order in which the building blocks are arranged that is characteristic of a particular DNA or protein molecule, and is responsible for its properties and how it functions.

## DNA

In DNA, there are four building blocks called bases or nucleotides. The four bases are usually represented by their initials—*A, C, G,* and *T*—so that a DNA sequence looks something like *TACCTGA*.[2] DNA generally exists in double strands because of the propensity for the bases to form hydrogen bonds with each other in a highly specific way: *A* with *T,* and *C* with *G.* Because of these complementary relations, given one strand, it is easy to recover the complementary strand. This aspect is of course essential during cell division when a cell must replicate its DNA. The normal conformation of double-stranded DNA is the famous double helix discovered by Watson and Crick in 1952.

DNA is the support of the genetic information required to build an entire organism, which includes the genes and their associated regulatory regions. A gene is a stretch of DNA that codes for a particular protein.[3] RNA, a class of molecules similar to DNA but single-stranded, is used, among other things, as an intermediary step in the translation of DNA into proteins. In some viruses, genetic information is stored using RNA sequences instead of DNA.[4]

## Proteins

A typical mammalian cell contains about 1 billion protein molecules. Proteins carry most of the operations of life at the molecular level. Some serve as building blocks for the cell, and others act as enzymes, accelerating chemical reactions. Proteins are continually being destroyed and replaced, so new ones must find their way to their proper posts. Proteins interact with each other, with DNA, with RNA, and with many simpler molecules. Hemoglobin, for instance, is a protein that binds and carries oxygen molecules in the blood. A number of proteins play a structural role. Collagen is a strong, fibrous protein that is an essential constituent of skin, hair, and bone tissues. Proteins also mediate many signals within and between cells, as well as between cells and the outside world. G-protein-coupled receptors, for instance, are a very large family of receptor proteins that play an essential role in signaling, for instance in the nervous system. They are embedded in the membrane of neurons and mediate a vast array of signals, from chemical (neurotransmitters, hormones, odorants) to light (photons). Most proteins, however, are enzymes and play a catalytic role. They accelerate, often by several orders of magnitude, all the complex chemical reactions involved with metabolic pathways and life. Without enzymes and at room temperature, such reactions would occur at rates too slow to be compatible with life's processes. Finally, certain proteins interact with DNA to proofread, edit, and repair DNA, to replicate or copy DNA, or to regulate transcription, that is to activate and suppress specific genes at specific times.

Protein sequences are made up of twenty different building blocks called amino acids.[5] The genetic code is a lookup table that specifies how the linear DNA sequence of a gene with its *A, C, G, T* alphabet is to be translated into the linear sequence of a protein made out of its twenty-amino-acid alphabet. Remarkably, the genetic code is the same for essentially all organisms.[6] Each triplet of DNA bases is translated into one of the amino acids. There are sixty-four possible triplets but only twenty amino acids, so the genetic code must be redundant. Three of the triplets correspond to stop signals used to mark the end of a gene. Proteins have a typical length of about 300 to 400 amino acids, equivalent to a gene length of about 1000 DNA letters.

## Genomes

If it were unwound and stretched out, all the DNA contained in the nucleus of a human cell would reach a length of several meters—more than human height. Thus DNA is tightly packed in the nucleus, which is only a few microns in radius. This is achieved through several levels of packing organization. At the first level beyond the double helix, DNA combines with certain proteins to form chromatin fibers.[7] At the highest level of spatial organization, DNA is split into units called chromosomes. Each human

Table 3.1 Typical genome sizes

| Organism | Number of DNA bases | Number of genes |
|---|---|---|
| Virus | 10–100,000 | 10 |
| Bacteria | 5 million | 4000 |
| Single cell | 15 million | 6000 |
| Simple animal or plant | 100 million | 20,000 |
| Human | 3 billion | 40,000 |

has twenty-three pairs of chromosomes. In each pair, one chromosome is contributed by the father and one by the mother; the two chromosomes are said to be homologous. The complete genetic sequence of an organism is called the genome.

Evolution has built organisms that vary in size over twenty-one orders of magnitude, from microbes ($10^{-13}$ gm) to whales ($10^{8}$ gm). It is useful to have an idea of genome sizes across different forms and levels of complexity of life (table 3.1). At one end of the spectrum, viruses have the smallest genomes. The AIDS virus, for instance, has a genome containing approximately 10,000 DNA bases including 9 genes (figure 3.2).[8] Single-cell organisms such as bacteria have genomes with 4000 genes or so.[9] More complex single-cell organisms have a slightly larger number of genes. The genome of yeast contains approximately 6000 genes. Thus a cell can be viewed as a complex system with 6000 different components or proteins. In the case of simple multicellular plants or animals, such as the little worm called nematode, the genome size reaches 97 million base pairs with over 19,000 genes. The nematode is just 1 mm long and has just under 1000 cells. At least 40% of nematode genes match those found in other animals. One of the most important discoveries of modern biology is the remarkable conservation of genes and developmental processes across different living species (figure 3.3). The fly's genome contains 180 million base pairs and approximately 14,000 genes, slightly over twice the number in yeast but significantly less than the worm. This is somewhat surprising, given the complexity of flies, and is only partially explained by the fact that many of its genes can be read in several forms (alternative splicing), giving rise to several proteins. In humans, the genome is close to 3 billion base pairs, with a total number of genes in the 40,000 range.[10] On average, we have

tggaagggctaattcactcccaacgaggacaagatatccttgatctgtggatctaccacacacaaggctacttccctgattggcagaactacacaccaggaccagggatcagatatccactgacctttggatggtgctacaagc
tagtaccagttgagccagagaagttagaagaagccaacaaaggagagaacaccagcttattacaccctgtgagcctgcatggaatggatgacccggagagagaagtgttagagtggaggtttgacagccgcctagcatttc
atcacgtagcccgagagctgcatccggagtacttcaagaactgctgacatcgagcttgctacaagggactttccgctggggactttccagggaggcgtggcctgggcgggactggggagtggcgagccctcagatgctgc
atataagcagctgctttttgcctgtactgggtctctctggttagaccagatctgagcctgggagctctctggctaactagggaacccactgcttaagcctcaataaagcttgccttgagtgcttcaagtagtgtgtgcccgtctgttgt
gtgactctggtaactagagatccctcagacccttttagtcagtgtggaaaatctctagcagtggcgcccgaacagggacttgaaagcgaaagggaaaccagaggagctctctcgacgcaggactcggcttgctgaagcgcg
cacggcaagaggcgaggggaggcgactggtgagtacgccaaaaattttgactagcggaggctagaaggagagagatgggtgcgagagcgtcagtattaagcgggggagaattagatcgatgggaaaaaattcggttaa
ggccagggggaaagaaaaaatataaattaaaacatatagtatgggcaagcagggagctagaacgattcgcagttaatcctggcctgttagaaacatcagaaggctgtagacaaatactgggacagctacaaccatcccttca
gacaggatcagaagaacttagatcattatataatacagtagcaaccctctattgtgtgcatcaaaggatagagataaaagacaccaaggaagctttagacaagatagaggaagagcaaaacaaaagtaagaaaaaagcacag
caagcagcagctgacacaggacacagcagccaggtcagccaaaattaccctatagtgcagaacatccaggggcaaatggtacatcaggccatatcacctagaactttaaatgcatgggtaaaagtagtagaagagaaggct
ttcagcccagaagtaatacccatgtttcagcattatcagaaggagccaccccacaagatttaaacaccatgctaaacacagtggggggacatcaagcagccatgcaaatgttaaaagagaccatcaatgaggaagctgcag
aatgggatagagtgcatccagtgcatgcagggcctatcgcaccaggccagatgagagaaccaaggggaagtgacatagcaggaactactagtacccttcaggaacaaataggatggatgacaaataatccacctatccca
gtaggagaaatttataaaagatggataatcctgggattaaataaaatagtaagaatgtatagccctaccagcattctggacataagacaaggaccaaaagaaccttttagagactatgtagaccggttctataaaactctaagagc
cgagcaagcttcacaggaggtaaaaaattggatgacagaaaccttgttggtccaaaatgcgaacccagattgtaagactattttaaaagcattgggaccagcggctacactagaagaaatgatgacagcatgtcagggagtag
gaggacccggccataaggcaagagttttggctgaagcaatgagccaagtaacaaattcagctaccataatgatgcagagaggcaattttaggaaccaaagaaagattgttaagtgtttcaattgtggcaaagaagggcacata
gccagaaattgcagggcccctaggaaaaagggctgttggaaatgtggaaaggaaggacaccaaatgaaagattgtactgagagacaggctaattttttagggaagatctggccttcctacaagggaaggccagggaatttc
ttcagagcagaccagagccaacagccccaccatttcttcagagcagaccagagccaacagccccaccagaagagagcttcaggtctggggtagagacaacaactccctctcagaagcaggagccgatagacaaggaact
gtatcctttaacttccctcagatcactctttggcaacgacccctcgtcacaataaagataggggggcaactaaaggaagctctattagatacaggagcagatgatacagtattagaagaaatgagtttgccaggaagatggaaac
caaaaatgatagggggaattggaggttttatcaaagtaagacagtatgatcagatactcatagaaatctgtggacataaagctataggtacagtattagtaggacctacacctgtcaacataattggaagaaatctgttgactctga
ttggttgcactttaaattttcccattagccctattggaactgtaccagtaaaattaaagccaggaatggatggcccaaaagttaaacaatggccattgacagaagaaaaaataaaagcattagtagaaatttgtacagagatggaaa
aggaagggaaaatttcaaaaattgggcctgaaaatccatacaatactccagtatttgccataaagaaaaaagacagtactaaatggagaaaattagtagatttcagagaacttaataagagaactcaagacttctgggaagttca
attaggaataccacatcccgcagggttaaaaaagaaaaaatcagtaacagtactggatgtgggtgatgcatatttttcagttcccttagatgaagacttcaggaagtatactgcatttaccatacctagtataaataatgagacacc
agggattagatatcagtacaatgtgcttccacagggatggaaaggatcaccagcaatattccaaagtagcatgacaaaaatcttagagccttttaaaaaacaaaatccagacatagttatctatcaatacatggatgatttgtatgta
ggatctgacttagaaatagggcagcatagaacaaaaatagaggagctgagacaacatctgttgaggtggggatttaccacaccagacaaaaaacatcagaaagaacctccattcctttggatgggttatgaactccatcctgat
aaatggacagtacagcctatagtgctgccagaaaaagacagctggactgtcaatgacatacagaagttagtggggaaattgaattgggcaagtcagatttacccagggattaaagtaaggcaattatgtaaactccttagagga
accaaagcactaacagaagtaataccactaacagaagaagcagagctagaactggcagaaaacagagagattctaaaagaaccagtacatggagtgtattatgacccatcaaaagacttaatagcagaaatacagaagcag
gggcaaggccaatggacatatcaaatttatcaagagccatttaaaaatctgaaaacaggaaaatatgcaagaatgaggggtgcccacactaatgatgtaaaacaattaacagaggcagtgcaaaaaataaccacagaaagca
tagtaatatggggaaagactcctaaatttaaactgcccatacaaaaggaaacatgggaaacatggtggacagagtattggcaagccacctggattcctgagtgggagtttgtcaatacccctcccttagtgaaactatggtacca
gttagagaaagaacccatagtaggagcagaaactttctatgtagatggggcagctaacagggagactagattaggaaaagcaggatatgttactaacaaaggaagacaaaaggttgtccccctaactaacacaacaaatcag
aagactgagttacaagcaatttatctagctttgcaggattcgggattagaagtaaacatagtaacagactcacaatatgcattaggaatcattcaagcacaaccagatcaaagtgaatcagagttagtcaatcaaataatagagca
gttaataaaaaaggaaaaggtctatctggcatgggtaccagcacacaaaggaattggaggaaatgaacaagtagataaattagtcagtgctggaatcaggaaaatactatttttagatggaatagataaggcccaagatgaaca
tgagaaatatcacagtaattggagagcaatggctagtgattttaacctgccacctgtagtagcaaaagaaatagtagccagctgtgataaatgtcagctaaaaggagaagccatgcatggacaagtagactgtagtccaggaat
atggcaactagattgtacacatttagaaggaaaagttatcctggtagcagttcatgtagccagtggatatatagaagcagaagttattccagcagaaacagggcaggaaacagcatattttcttttaaaattagcaggaagatggc
cagtaaaaacaatacatacagacaatggcagcaatttcaccagtactacggttaaggccgcctgttggtgggcggggatcaagcaggaatttggaattccctacaatcctcaaagtcaaggagtagtagaatctatgaataaag
aattaaagaaaattataggacaggtaagagatcaggctgaacatcttaagacagcagtacaaatggcagtattcatccacaattttaaaagaaaaggggggattggggggtacagtgcaggggaaagaatagtagacataata
gcaacagacatacaaactaaagaattacaaaaacaaattacaaaaattcaaaattttcgggtttattacagggacagcagaaatccactttggaaaggaccagcaaagctcctctggaaaggtgaaggggcagtagtaataca
agataatagtgacataaaagtagtgccaagaagaaaagcaaagatcattagggattatggaaaacagatggcaggtgatgattgtgtggcaagtagacaggatgaggattagaacatggaaaagtttagtaaaacaccatatg
tatgtttcagggaaagctaggggatggttttatagacatcactatgaaagccctcatccaagaataagttcagaagtacacatcccactaggggatgctagattggtaataacaacatattggggtctgcatacaggagaaagag
actggcatttgggtcagggagtctccatagaatggaggaaaaagagatatagcacacaagtagaccctgaactagcagaccaactaattcatctgtattactttgactgtttttcagactctgctataagaaaggccttattaggac
acatagttagccctaggtgtgaatatcaagcaggacataacaaggtaggatctctacaatacttggcactagcagcattaataacaccaaaaaagataaagccacctttgcctagtgttacgaaactgacagaggatagatgga
acaagccccagaagaccaagggccacagagggagccacacaatgaatggacactagagcttttagaggagcttaagaatgaagctgttagacattttcctaggatttagctccatggcttaggacaacatatctatgaaactta
tggggatacttgggcaggagtggaagccataataagaattctgcaacaactgctgtttatccattttcagaattgggtgtcgacatagcagaataggcgttactcgacagaggagagcaagaaatggagccagtagatcctaga
ctagagccctggaagcatccaggaagtcagcctaaaactgcttgtaccaattgctattgtaaaaagtgttgctttcattgccaagtttgtttcataacaaaagccttaggcatctcctatggcaggaagaagcggagacagcgacg
aagacctcctcaaggcagtcagactcatcaagtttctctatcaaagcagtaagtagtacatgtaatgcaacctatacaaatagcaatagtagcattagtagtagcaataataatagcaatagttgtgtggtccatagtaatcatagaa
tataggaaaatattaagacaaagaaaaatagacaggttaattgatagactaatagaaagagcagaagacagtggcaatgagagtgaaggaaaaatatcagcacttgtggagatgggggtggagatggggcaccatgctcctt
gggatgttgatgatctgtagtgctacagaaaaattgtgggtcacagtctattatggggtacctgtgtggaaggaagcaaccaccactctattttgtgcatcagatgctaaagcatatgatacagaggtacataatgtttgggccaca
catgcctgtgtacccacagaccccaacccacaagaagtagtattggtaaatgtgacagaaaattttaacatgtggaaaaatgacatggtagaacagatgcatgaggatataatcagtttatgggatcaaagcctaaagccatgtg
taaaattaaccccactctgtgttagtttaaagtgcactgatttaaagaatgatactaataccaatagtagtagcgggagaatgataatggaaaaaggagagataaaaaactgctctttcaatatcagcacaagcaaaagaggtaag
gtgcagaaagaatatgcatttttttataaacttgatataataccaatagataatgatactaccagctatacgttgacaagttgtaacacctcagtcattacacaggcctgtccaaaggtatcctttgagccaattcccatacattattgtg
ccccggctggttttgcgattctaaaatgtaataataagacgttcaatggaacaggaccatgtacaaatgtcagcacagtacaatgtacacatggaattaggccagtagtatcaactcaactgctgttaaatggcagtctagcagaa
gaagaggtagtaattagatctgccaatttcacagacaatgctaaaaccataatagtacagctgaaccaatctgtagaaattaattgtacaagacccaacaacaatacaagaaaaagaatccgtatccagaggggaccagggag
agcatttgttacaataggaaaaataggaaatatgagacaagcacattgtaacattagtagagcaaaatggaatgccactttaaaacagatagctagcaaattaagagaacaatttggaaataataaaacaataatctttaagcaat
cctcaggaggggacccagaaattgtaacgcacagttttaattgtggaggggaatttttctactgtaattcaacacaactgtttaatagtacttggtttaatagtacttggagtactgaagggtcaaataacactgaaggaagtgacac
aatcacactcccatgcagaataaaacaatttataaacatgtggcaggaagtaggaaaagcaatgtatgcccctcccatcagcggacaaattagatgttcatcaaatattacagggctgctattaacaagagatggtggtaataac
aacaatgagtccgagatcttcagacctggaggaggagatatgagggacaattggagaagtgaattatataaatataaggtagtaaaaattgaaccattaggagtagcacccaccaaggcaaagagaagagtggtgcagaga
gaaaaaagagcagtgggaataggagctttgttccttgggttcttgggagcagcaggaagcactatgggctgcacgtcaatgacgctgacggtacaggccagacaattattgtctggtatagtgcagcagcagaacaatttgct
gagggctattgaggcgcaacagcatctgttgcaactcacagtctggggcatcaagcagctccaggcaagaatcctggctgtggaaagatacctaaaggatcaacagctcctggggatttggggttgctctggaaaactcattt
gcaccactgctgtgccttggaatgctagttggagtaataaatctctggaacagatttggaataacatgacctggatggagtgggacagagaaattaacaattacacaagcttaatacactccttaattgaagaatcgcaaaacca
gcaagaaaagaatgaacaagaattattggaattagataaatgggcaagtttgtggaattggtttaacataacaaattggctgtggtatataaaattattcataatgatagtaggaggcttggtaggtttaagaatagtttttgctgtactt
tctgtagtgaatagagttaggcagggatattcaccattatcgtttcagacccacctcccaaccccgaggggacccgacaggcccgaaggaatagaagaagaaggtggagagagagacagagacagatccattcgattagtg
aacggatccttagcacttatttgggacgatctgcggacgctgtgcctcttcagctaccaccgcttgagagacttactcttgattgtgacgaggattgtggaacttctgggacgcagggggtgggaagccctcaaatattggtgga
Atctcctacaatattggagtcaggagctaaagaatagtgctgttagcttgctcaatgccacagctatagcagtagctgaggggacagatagggttatagaagtagtacaaggagcttgtagagctattcgccacatacctagaa
gaataagacagggcttggaaaggattttgctataagatgggtggcaagtggtcaaaaagtagtgtgattggatggcctgctgtaagggaaagaatgagacgagctgagccagcagcagatggggtgggagcagcatctcg
agacctagaaaaacatggagcaatcacaagtagcaacacagcagctaacaatgctgcttgtgcctggctagaagcacaagaggaggagaaggtgggttttccagtcacacctcaggtacctttaagaccaatgacttacaag
gcagctgtagatcttagccactttttaaaagaaaaggggggactggaagggctaattcactcccaacgaggacaagatatccttgatctgtggatctaccacacacaaggctacttccctgattggcagaactacacaccagga
ccagggatcagatatccactgacctttggatggtgctacaagctagtaccagttgagccagagaagttagaagaagccaacaaaggagagaacaccagcttattacaccctgtgagcctgcatggaatggatgacccggag
agagaagtgttagagtggaggtttgacagccgcctagcatttcatcacgtagcccgagagctgcatccggagtacttcaagaactgctgacatcgagcttgctacaagggactttccgctggggactttccagggaggcgtgg
cctgggcgggactggggagtggcgagccctcagatgctgcatataagcagctgctttttgcctgtactgggtctctctggttagaccagatctgagcctgggagctctctggctaactagggaacccactgcttaagcctcaat
aaagcttgccttgagtgcttcaagtagtgtgtgcccgtctgttgtgtgactctggtaactagagatccctcagacccttttagtcagtgtggaaaatctctagca

Figure 3.2 The complete genome of an HIV virus, containing 9752 letters and 9 genes. (From GenBank).

Figure 3.3 Large, plump tomatoes can evolve from wild, blueberry-size tomatoes. A single gene expressed early in the plant's floral development appears to be largely responsible for the plumping characteristic. Computer analysis reveals that the corresponding protein appears to be similar to a human protein involved in cancer, suggesting common cell proliferation mechanisms leading to large, edible fruits in plants and cancer susceptibility in humans (Frary et al. 2000). In the era of genomics, such unexpected connections between divergent organisms are becoming more and more common. (Courtesy of Kent Loeffler and the Department of Plant Pathology, Cornell University.)

about 130 million DNA base pairs in each chromosome. Curiously, humans are not the organisms with the largest genome. Some plants in particular, as well as certain amphibians and fish, have much larger genomes, although not as many genes. The genome of the lily *Fritallaria* is 40 times the size of the human genome. Overall, evolution does not build more complex forms of life by greatly expanding the number of genes, but by creating novel combinations and especially novel interactions (i.e., different ways of modulating the expression of the genes).

Thus viruses have very compact genomes, with genes that are continuous or even overlapping. In higher organisms, gene sequences are typically separated by long stretches of nonprotein-coding DNA. In fact, gene coding sequences often are broken up into alternating coding regions, called exons, interrupted by noncoding intervening regions called introns. Exons have a typical length in the range of 100 to 200 base pairs; introns are highly variable in length and often cover several thousand base pairs. Thus most DNA in higher organisms does not correspond to genes and does not have any coding functions.

Perhaps the most striking feature of genomic information is the fact that it is both discrete and finite: with sufficient effort it can be obtained entirely. The three gigabytes of the human genome easily fit on the hard drive of today's personal computers and are roughly equivalent to the size of a very large encyclopedia of 300 volumes, each containing 2000 pages with 5000 characters on each page—an encyclopedia that contains the complete blueprint of a human being.

To have a sense for the complexity of human gene finding, an analogy can be useful. The total number of chromosomes in a human cell is comparable to the number of states in the Union. Imagine that each state happens to be a long, thin strip of land like California, with a typical length of 1000 kilometers. Thus, if a chromosome were a state, each DNA nucleotide would correspond very roughly to 1 centimeter.[11] A typical gene would spread out over a few hundred meters, but its coding portion would be only about 10 meters long, broken up into exons of 1 meter. The Human Genome Project would turn into a big census of all the major highways in the United States, centimeter by centimeter, with the additional goal of detecting all the billboards (genes), and later to compare, classify, and understand the message on each billboard. This would have to be achieved using a pool of very slow golf carts, with an autonomy of about 400 letters—or 4 meters—each! And because of census errors, ten golf carts would have to check each centimeter of highway.

## The Stadium-Size Cell

*Any living cell carries with it the experience of a billion years of experimentation by its ancestors.*

—Max Delbrück, physicist/biologist

It is also very useful to try to put DNA and proteins in the perspective of a cell. The amazing beauty and complexity of a cell is not always easy to grasp because of the very small sizes involved. Atom sizes are measured in Ångströms (1 Å = $10^{-10}$ m, a tenth of a billionth of a meter). Cells have typical radius of 10 to 30 microns (1 μ = $10^{-6}$ m, a millionth of a meter), and a meter is roughly the same thing as a yard (in case you are not used to the metric system). The volume occupied by a protein can be estimated by noticing that it is made of many amino acids. An amino acid consists of a few dozen atoms and typically occupies a volume of 100 cubic Å. Thus the following list gives the typical length scales that are important at the cellular level.[12]

- atom length = 1 Å
- diameter of DNA double helix = 10 Å
- volume of single amino acid = 100 Å$^3$ [roughly 5 Å × 5 Å × 5 Å]
- volume of single protein = $10^5$ Å$^3$ [roughly 50 Å × 50 Å × 50 Å]
- thickness of cell membrane = 60–80 Å
- radius of nucleus = 3–10 μ
- radius of cell = 10–30 μ

Because such minute sizes are difficult to visualize, we can again use an analogy by blowing everything up by a factor of 10 million ($10^7$). After zooming down to California-size chromosomes, we can zoom out a little and take a look at Rose Bowl-stadium-size cells.

- atom length = 1 millimeter = the size of a sand grain
- diameter of DNA double helix = 1 centimeter
- volume of amino acid = 100 mm$^3$ [roughly 5 mm × 5 mm × 5 mm] = the size of a pearl

- volume of protein = 100 $cm^3$ [roughly 5 cm × 5 cm × 5cm] = the size of a tennis ball
- thickness of cell membrane = 6–8 cm
- radius of nucleus = 30–100 meters = the size of a large tennis court or a football field
- radius of cell = 100–300 meters = the size of a large stadium

Thus, we can think of a cell as a large stadium continuously surrounded by a relatively thin wall as well as a dome. The nucleus is like the football field. The players in the center are the chromosomes. Each chromosome consists of a DNA thread with a diameter of 1 centimeter and a length of 1000 kilometers. It is worth noticing that the blowup factor of $10^7$ is the same as the one used above, except that the DNA was stretched out like the state of California. Stretched out, the thread contained in all the chromosomes would go once around the Earth. Now in the stadium it is folded up compactly in a big spaghetti mess. The folded spaghetti mess, however, is much bigger than a football player because it forms a sort of 10-meter-high tower.

The stadium is filled with sand grains of many different sizes and types, moving in all directions, ranging from water molecules, to sodium, potassium, calcium, and chloride ions.[13] In the stadium, proteins come in many shapes and sizes, but typically have the dimensions of a tennis ball. If we assign a different color to each protein type, or each gene, in a human cell, we will have on the order of 40,000 different colors—roughly the number of spectators. Each protein is represented by a variable number of replicas. Each cell type at a given time has a different overall blend of colors that depends on the genes that are being expressed. The precise code for manufacturing each tennis ball is buried in the spaghetti mess on the football field, but the proteins are manufactured in the stadium outside the football field. Overall, proteins are extremely busy in the stadium as they continually bind and interact with each other, with the spaghetti mess, and with various sand grains.

Somehow proteins must find their way to the region of their activity: the football field (nucleus), the rest of the stadium (cytoplasm), the wall around the stadium (membrane), or even the external world in the case of secreted proteins. They are what keeps the stadium functioning, by generating energy, removing waste, exchanging food and other signals with the external world, producing other tennis balls, fighting enemies, and so on. Some of them work with the spaghetti mess, repairing it, looking for the instructions for a particular type of tennis ball, and starting the production process. From time to time, proteins take care of the very complex events by which an entire stadium is precisely duplicated into two stadiums and each spaghetti mess into two identical spaghetti messes. In chapter 9 we will come back to the stadium analogy for neurons, the cells responsible for processing in our brains.

## Evolution

*Evolution is true—and the truth can only make us free.*

—Stephen Jay Gould, paleontologist

Genomes are not static entities, but dynamic ones. They are constantly undergoing a variety of changes, from small ones such as mutations to larger ones such as recombinations. Recombination happens during meiosis, the cell division that occurs during the formation of egg and sperm cells and gives them half the number of chromosomes contained in other body cells. Since egg and sperm unite at fertilization, each must carry only half the total number of chromosomes. During meiosis, each chromosome in a pair is segregated to a daughter sex cell independent of all the others. The total number of such possible combinations is $2^{23}$, over 8 million, for each sex cell. Thus, as a result of chromosome assortment alone, the number of genetically different possible descendants of a couple is $2^{46}$, close to 7 trillion. In fact, the number is even larger because chromosomes in a pair also tend to swap genetic material with each other in what is termed crossover.[14] Modifications of genetic molecules and selective pressure from the environment are the two basic components of evolution. Because genomes are constantly changing, the taxonomy of living things cannot be rigid, and boundaries between its objects cannot be sharply defined, including the definition of *Homo sapiens*. It is no surprise that such classifications are constantly being revised and updated by biologists.

Using a variety of methods, including radiometric dating of fossils and the comparison of genomic sequences, it is possible to estimate the dates of a number of evolutionary events. The following list provides a sense of the cosmic, animal, and human scales of evolution.

- origin of universe (Big Bang) = 10–15 billion years
- origin of solar system and Earth = 4.5 billion years
- first living cells = 3.5 billion years
- first multicellular organisms = 1 billion years
- invasion of dry land by plants = 400 million years
- first mammals and dinosaurs = 250 million years
- extinction of dinosaurs = 60 million years
- emergence of *Homo* = 2 million years
- human language = 50,000 years
- human societies = 10,000 years
- written language = 5000 years
- scientific method = 400 years
- industrial revolution = 200 years
- DNA structure and first computers = 50 years

It is not easy for our minds to comprehend some of the larger numbers. For the sake of comparison, consider continental drift, which occurs at the very slow rate of the order of 1 centimeter per year, or 1 meter per century. For a continent to drift the length of an ocean (5000 kilometers), it takes 500 million years. In fact, the five modern continents drifted away from a single land mass, *Pangaea,* about 200 million years ago. Thus continental drift occurs almost at a fast pace with respect to the history of life on Earth, and it is no wonder that it played an important role in the differentiation of species.

Finally, one may also want to have a rough idea of the total number of humans who have lived throughout our short evolution. The total number of living humans in the year 2000 is on the order of 6 billion. Using a very crude evolutionary model, one can easily get an estimate of about 10 to 100 billion (see appendix A).

## Genome Sequencing

With current sequencing technology, relatively long stretches of DNA can be sequenced at once. By sequencing and assembling many overlapping fragments, it is possible to sequence even large genomes completely. This process can be costly and tedious, but it has no secrets. Today any organism can be sequenced in principle; it is only a question of time, effort, and money. Furthermore, the speed and cost of sequencing technologies are undergoing constant improvements. It took a month of work to sequence a single DNA base pair around 1970. Celera Genomics, the largest sequencing facility in the world today, can produce on the order of 1 billion base pairs in the same amount of time.

The choice of genomes being sequenced so far has been dictated by a number of considerations including historical, experimental convenience, and availability of complementary genetic and physiological information. In addition to viruses, organisms that have played a prototypical role in biological studies and sequencing include (with their exotic Latin names): *Escherichia coli* (bacterium), *Saccharomyces cerevisiae* (yeast), *Caenorhabditis elegans* (worm), *Arabidopsis thaliana* (plant), *Drosophila megalonaster* (fly), *Mus musculus* (mouse), *Sus scrofa* (pig), and of course *Homo sapiens* (man). By the end of the twentieth century, several dozen genomes had been entirely sequenced, including many viruses and bacteria, yeast, the worm, and the fly. The worm's genome, the first genome of a multicellular organism, was completed in December 1998. The fly's genome was completed in early 2000, closely followed by the rice genome. Many other genomes, such as mouse, have been partially sequenced and certainly will be completed during the first years of the third millennium.

The Human Genome Project, outlined in 1986 and officially launched in 1990, is probably the largest and farthest-reaching scientific project ever undertaken by humans, with profound impacts on biology, medicine, industry, and society. It is a collective, accelerating, scientific undertaking involving thousands of scientists collaborating at various levels and culminating in a race between universities and public laboratories, on the one hand, and commercial organizations, such as Celera Genomics, on the other. The first chromosome (chromosome 22) was completed in late 1999, and a rough version of the entire genome in 2000, jointly announced by Celera and the public Human Genome Project consortium. In 1953, Watson and Crick discovered DNA's fundamental structure. In less than half a century we will have progressed from the double helix to the complete genetic blueprint of humankind.

The project provides a single prototypical human genome sequence based on DNA from a small number of individuals, but it is also important to sequence DNA across large populations of individuals to understand genetic variability and, for instance, its impact on both the occurrence of diseases and the effectiveness of therapies. Many complementary projects of this kind are under way, such as the study of SNPs (single-nucleotide polymorphisms). These are specific one-letter positions along the genome, with a density of roughly 1 every 1000 bases, where a substantial fraction of the genetic variability between humans is concentrated. Thus any two randomly selected humans differ on average in less than 0.5% of their genomes. It is perhaps surprising that variability within chimpanzees seems to be higher. In any case, these variations can have a major impact on how humans respond to diseases, environment, drugs, and therapies.

It is well known, for instance, that the distribution of responses to any given drug is in general fairly broad. In other words, for any given drug a subset of individuals with a particular genetic makeup do not respond well to it and would therefore benefit from a different therapy. In 1999, ten large pharmaceutical companies and a British charity announced the creation and 45 million dollar funding of a consortium—named TSC—for a public database of human SNPs, to be constructed in collaboration with a number of academic laboratories. In addition to the Wellcome Trust of Britain, the sponsors are Bayer Group AG, Bristol-Myers Squibb Co., Glaxo Wellcome PLC, Hoechst Marion Roussel AG, Monsanto Co., Novartis AG, Pfizer Inc., Roche Holding Ltd., SmithKline Beecham PLC, and Zeneca Group PLC. Well over a million SNPs had been identified by the end of 2000. There is little doubt that the molecular medicine of the future will take into account the precise genetic composition of each individual.

To provide further sequencing incentives that would justify the corresponding costs, genes have been made patentable, a significant deviation from the original dichotomy between patentable man-made objects and nonpatentable objects resulting from the work of nature.[15] Indeed, many organizations, from commercial companies to universities, are rushing to develop and complete a variety of large sequencing projects. As a result, genetic databases are growing very rapidly. At the end of the 1990s, public databases were doubling every sixteen months or so. Remarkably, this is roughly the exponential pace followed by the computing power, or the price, of our computers (see Moore's law in chapter 8). An example of a large commercial sequencing project we have already seen was announced in 1998: a "parallel" human genome sequencing project by Craig Venter of Celera Genomics Corporation, partly owned by Perkin–Elmer ABI. The project was started with approximately 300 ABI Prism 3700 sequencing machines.[16] The stated initial goal was to patent a few hundred human genes of great medical significance.

## Beyond Genome Sequencing

*The biology of the future is going to be the study of how networks of genes and proteins operate in the context of complex living systems.*

—Leroy Hood, biologist

Whereas sequencing an entire genome is a useful step, by itself it only creates massive amounts of raw data, akin to the list of spare parts of a Boeing 747, that have to be managed and further processed to extract useful information and understand how the pieces work together. This is in part the domain of bioinformatics. Some of the most immediate tasks in bioinformatics are to find the genes, the lexicon of their regulatory regions, and the structure and function of the corresponding proteins, as well as the reconstruction of phylogenetic trees.[17] Finding genes and regulatory regions is not a trivial task because the corresponding statistical patterns are highly variable and genes are in general discontinuous. Furthermore, they represent only a minute fraction of the total amount of DNA in higher organisms. Regulatory regions, such as promoters, typically correspond to regions upstream or downstream of a gene. There, proteins can bind and interact with DNA in complex ways—for instance, to initiate or inhibit the transcription of a gene, the first step toward its expression as a protein.[18] Knowledge of a gene sequence translates into knowledge of the corresponding protein sequence by the genetic code. But even there, additional technical problems can arise—for instance, if a portion of the sequence is missing, if the reading frame is not known, or if additional transitory elements that are not part of the final protein product are included.[19]

Proteins are three-dimensional molecular machines. Determining the structure and function of a protein from its sequence alone is a complex task for a number of reasons. We still do not entirely understand how proteins fold in three dimensions. Furthermore, proteins often undergo a number of additional chemical transformations, such as phosphorylation and glycosylation, after their translations or their folding. Although we may have only 40,000 genes, combinations and modifications may give rise to 1 to 10 million chemically different proteins in a cell. Finally, the function of a protein is a multifaceted notion that covers many different attributes. These include notions as diverse as the cellular location of the mature protein (secreted, membrane, cytoplasm, nucleus), the proteins and other substrates it interacts with and the mode of interaction, the metabolic pathways and protein networks it is associated with, the corresponding gene regulatory network, and the phenotypic effects and dysfunctions. These issues, however, can be addressed through a number of experimental and computer techniques, ranging from X-ray crystallography to gene and protein chips, to sequence comparison and machine learning.

More generally, scientists are addressing all these questions simultaneously, using multiple computer and experimental methods, across different sequences and organisms for which variable amounts of knowledge are available during the current transitory phase of discovery. The trend in fact is toward genomics and proteomics, the study of large numbers of genes and proteins in parallel from different angles, such as functional, structural, regulatory, and evolutionary genomics. Genes and proteins are for biology the equivalent of the periodic table for chemistry. The main goal now is to progress from the level of single molecules, genes, and proteins to the emergent system level, the level of gene networks or protein networks associated with gene regulation and metabolic pathways.

Thus the central questions of biology at the beginning of the third millennium are about understanding complex systems. For instance, over 300 genes are believed to be involved in the control of blood pressure in humans. How are the corresponding pathways organized? How does the entire organization of a single cell emerge from its DNA and a network of about 6000 proteins? How does the immune system work? How does an organism develop from a single cell to a complex assembly of $10^{15}$ cells? How do populations of interacting organisms and species evolve in an ecosystem? And, ultimately, how do complex mental functions emerge from the interaction of a trillion or so neurons in the brain?

We do not really know whether general principles of organization are used in biology across different levels, although there is a clear sense of hierarchical organization. For instance, a single mutation in the fly can result in the development of two pairs of wings: a small change at the top of a hierarchical developmental program can result in large changes in phenotype. But it is important to realize that the problems are finite and the general path to solve them is quite clear, although it may require a gigantic amount of brute force work. In fact, the question is more about finding clever shortcuts. There is little doubt, for instance, that in the relatively near future we will know all 6000 proteins of yeast and solve the structure and function of each of them. In particular, all existing proteins can be classified into families sharing common characteristics. The total number of families is remarkably manageable, estimated to be on the order of a few thousand. There are already many hundred families for which the basic structure and the function are relatively well understood. More generally, there appears to be a large but finite repertoire of statistical profiles or motifs out of which genes and proteins are built. In contrast, if one had asked Archimedes, Newton, or even somebody living in the early 1800s how to carry video at a distance of 1000 miles almost instantaneously, they would have been at a loss for where to start. Yet television has completely solved that problem in less than a century.

In humans, the total number of genes is very approximately on the order of 40,000—the number of people in a large, packed stadium. By early 1999, about 40% of the genes had been identified, but the function of only a much smaller fraction was known precisely. The expression of a given gene depends directly on a relatively small number of proteins, on the order of fifty.[20] A few proteins also play a major role by controlling the expression of many genes. This hierarchical regulatory circuitry is of course fundamental during development and throughout the life of an organism, yet only a few components of this circuit are known today, and progress on any component requires lengthy experiments. Methods exist, however, by which the entire set of genes or proteins expressed in a cell at a given time[21] can be monitored, and rapid progress in this area is continuing. One of these methods is the use of biochips or microarrays, such as DNA microarrays. Microarrays allow testing and visualizing the reaction of a number of DNA or protein sequences against a large array or library of other sequences or compounds. They provide a global snapshot of the gene or protein activity of a cell at a given time under a given set of conditions. For example, the single-stranded mRNA sequences associated with all the genes transcribed in a cell at a given time can be collected and distributed over an array containing a library of a large number of genes from a given organism. How the sequences pair up, or hybridize, can be visualized using fluorescence techniques that provide a snapshot of the global transcription activity of the cell.[22] Such snapshots can be used to study fundamental biological phenomena such as development or evolution, to determine the function of new genes, to infer the role individual genes or a group of genes may play in diseases, and to monitor the effects of drugs and other compounds on gene expression. One of the great promises of molecular medicine is the possibility of using such snapshots to provide diagnostic and therapeutic tools closely tailored to each individual's genetic makeup. Needless to say, these techniques and many others are in the process of profoundly changing biology, medicine, and drug discovery.

## Biotechnology

Besides sequencing and analysis of DNA, proteins, and genomes, many other operations are becoming possible through modern biotechnology. Biotechnology has begun to generate significant economic benefits, including production of valuable biological molecules and modified animals and plants for agriculture. It has also begun to generate biological systems that are closer and closer to science fiction. Today we routinely mutate and modify DNA and proteins. We control selective pressure and direct evolution in the laboratory. We transfer genetic material across species, even from animals to plants. We have expressed human genes in bacteria, yeast, mice, lambs, and cows—one way of producing large amounts of a given human protein, and possibly of a protein drug. Genetically modified bacteria, for instance, were used to produce insulin. In 1999, scientists at the British biopharmaceutical company PPL Therapeutics[23] (a company associated with the Roslin Institute in Scotland, where Dolly the sheep was cloned) were able for the first time to insert human genes into a pair of lambs to aid burn victims. The transgenic lambs carry a human gene that gives them the ability to produce human serum albumin, a protein that is often used in surgery and is essential for treating burn victims. The animals are expected to produce milk containing the serum, which will be extracted and used to create drugs. A patent application for this method has been filed.

In fact, not only single genes, but also an entire human chromosome with thousands of genes, have been inserted in mice. In a transgenic experiment published in 1997, Tomizuka and his collaborators succeeded in inserting human chromosome number 2 in mice embryos, using a relatively complex technique. They obtained a set of healthy mice with forty-one chromosomes, rather than the twenty pairs characteristic of normal mice. Chromosome 2 contains genes that are responsible for the synthesis of antibodies. The scientists were able to check that the extra human chromosome carried by the mice was functional by injecting the mice with antigens and showing that human antibodies were being secreted in addition to mice antibodies. Furthermore, the mice reproduced, and the extra chromosome was present in their progeny. These transgenic mice have many applications in medicine.

Even more exotic experiments have been performed, such as inserting a firefly gene into a tobacco plant to make it glow in the dark (see figure 3.1)—still unlikely to become the new wave for lighting cities at night. Likewise, a jellyfish gene has been used to produce a green mouse that glows in the dark.[24] And mouse genes involved in eye growth have been inserted into the genomes of fruit flies, resulting in the development of eye structures on the wings or legs of the flies.

To a certain extent, we can directly synthesize new DNA sequences and proteins from scratch. For instance, machines are now commercially available that synthesize DNA chains from nucleotide sequences typed on a keyboard. Current synthesizers can accurately make chains a few dozen nucleotides in length, the oligonucleotides. We are not yet able to synthesize entire genes and chromosomes because of the length and all the packaging involved, but we can modify and combine existing ones in many ways. Although a few lingering technical limitations on the precision and scope of such alterations remain, the boundaries are receding every day. We can direct evolution in the laboratory and certainly in the computer. In some cases, as with the influenza virus, we can begin to predict its future.

Gene therapies are being developed to treat some of the more than 4000 known genetic diseases. Many diseases result from the effects of combinations of genes, and some, such as breast cancer, can have several different forms associated with different genes. A few diseases, however, result from a single defective DNA region. The first one to be characterized at the molecular level is sickle-cell anemia. When the sickle gene is present in both homologous chromosomes, severe anemia results, characterized by red blood cells having a sickle shape. The defect is caused by a change in a single DNA base (*A* into *T*), resulting in a single amino acid difference in the corresponding hemoglobin protein. Two examples of over a dozen severe human degenerative diseases are myotonic dystrophy and Huntington's disease. All the disorders in this group result from a similar accident, the expansion of a triplet of DNA letter at a particular location in the genome. In healthy subjects the three DNA letters *CAG* are repeated six to thirty-five times in a row inside a specific gene located on chromosome 4. In people with Huntington's disease, the triplet is repeated a much larger number of times, sometimes over 100 times. Gene therapy aims, among other things, at correcting the negative effects of "defective" genes at the DNA level—for instance, by trying to introduce healthy versions of the genes inside the organism. We do not know how to cut out a given gene from a genome and systematically replace it with another one in *all* the cells of an adult multicellular organism. To affect all the cells genetically, we must do it at the first stages of development. But in many cases it is sufficient if the healthy gene or its product can be administered locally, where its presence matters. Although still in its infancy, gene therapy is an important direction for medicine in the new century.[25]

Although we are accustomed to the phrase "the human body," our bodies are genetically quite different. The differences are large enough that today a number of drugs which could benefit a fraction of the population cannot be approved because of side effects in a different fraction of the population. Current research is laying the foundation for a molecular medicine tuned to the DNA of an individual patient.

Going beyond medicine, because of their great chemical versatility, proteins are used in a number of industries, from pharmaceutical to food to paper. Organisms that live in extreme conditions, such as high pressure, temperature, or acidity, are being sequenced to find new proteins capable of functioning under those conditions. For instance, thermophilic bacteria, such as *Thermus aquaticum,* thrive in the boiling heat of geysers and geothermal vents at the bottom of the ocean. In particular, some of these bacteria engineer the protein DNA polymerase that can synthesize DNA at relatively high temperatures. This enzyme is widely used in laboratories to synthesize large amounts of a given DNA sequence in the polymerase chain reaction—the equivalent of DNA photocopying.[26]

We have engineered new animals and plants, such as the FlavrSavr tomato, developed by the biotechnology company Calgene, which can resist rotting. Plants are also being engineered that can cope with metals and improve human nutrition and human health. Recently, seven genes were successfully inserted in rice so that it contains beta-carotene, the precursor to vitamin A, as well as iron. Similarly, we have engineered plants that produce an insect-toxic protein, as well as plants resistant to common herbicides. Over 40% of the corn and 50% of the cotton planted in the United States in 1999 was genetically modified, substantially reducing the use of chemical pesticides. Overall, an estimated 28 million hectares of transgenic plants were grown worldwide in 1999, and the number is expected to increase rapidly in the coming years. We have turned bacteria into chemical reactors. We have started to grow or engineer artificial tissues and organs. We have cloned not only individual cells, but entire adult animals such as frogs, sheep, cows, pigs, mice, and monkeys.

## Genomes as Computer Programs

To a certain extent, it is useful to think of genomes in terms of a computer metaphor. A genome can be viewed as a formidable computer program written using the four-letter alphabet of DNA. This program is packaged in a special way, going from the double helix to the chromatin fibers and all the way up to the chromosomes. Packaging is important for execution. The computer unit that reads such a program and executes it is the cell. In a biological system, however, it is almost impossible to keep a clean separation between software and hardware. In fact, the execution of the software modifies the computer used to read it, and vice versa. DNA is a bit of both, and so are neurons and neuronal networks in the brain. Hence, we will often use the term "wetware" to denote biological software-hardware. It is in part because software and hardware are so interwoven and tuned to one another that biological systems have very large computational abilities and are difficult to reverse-engineer.

The human genome has been described as the most complex piece of software on Earth, although there is nothing preventing us from being able one day to design even more complex programs. It is well beyond detailed comprehension by a single human—but so are airplanes—and will require many teams of experts, many computers, and many decades (if not centuries) of hard work to reverse-engineer. Its complex, chemically mediated interactions with the hardware of the human body add layers upon layers of intricate inquiries, messages, commands, and feedback-induced countercommands issued by thousands upon thousands of components. Fortunately, although the numbers are large, they are finite; mapping complete biological systems will be feasible.

Executing the genomic program for the first time amounts to starting the development cycle of an organism. In passing, it is worth noticing that we inherit more than just half of our chromosomes from our mother: we inherit also the first egg cell, which comes with an array of nutrients, proteins, and organelles,[27] and the built-in capability to bootstrap the execution of the genetic instructions. Each one of us has gone through this astonishing program once. We have all grown from a single cell to an adult organism with trillions of cells. Through many precisely orchestrated cycles of cell division and differentiation, the program has created in each one of us over 250 cell types organized in complex interacting networks, tissues, and organs. The execution of the program does not stop with adulthood, but proceeds in each one of our cells throughout our lives. In each cell, the execution leads to the expression of a particular set of proteins that varies in time and depends on the internal state of the cell, as well as on its interactions with the environment, including other cells.

The foundation of the current molecular biotechnology revolution is precisely the ability to treat genomes as programs, and to be able to perform all the standard operations one performs on computer files: read, write, execute, copy, edit, cut-and-paste, insert, delete, replace, transfer, and so forth.[28] Perhaps even more so than for computers, issues of formatting and compatibility across genetic programs and genetic wetware are important. But at the most basic level, DNA and the genetic code provide a common universal language. We can perform several of these operations with a degree of flexibility and specificity, and on a scale which far exceeds that reached by nature during 3.5 billion years of evolution.

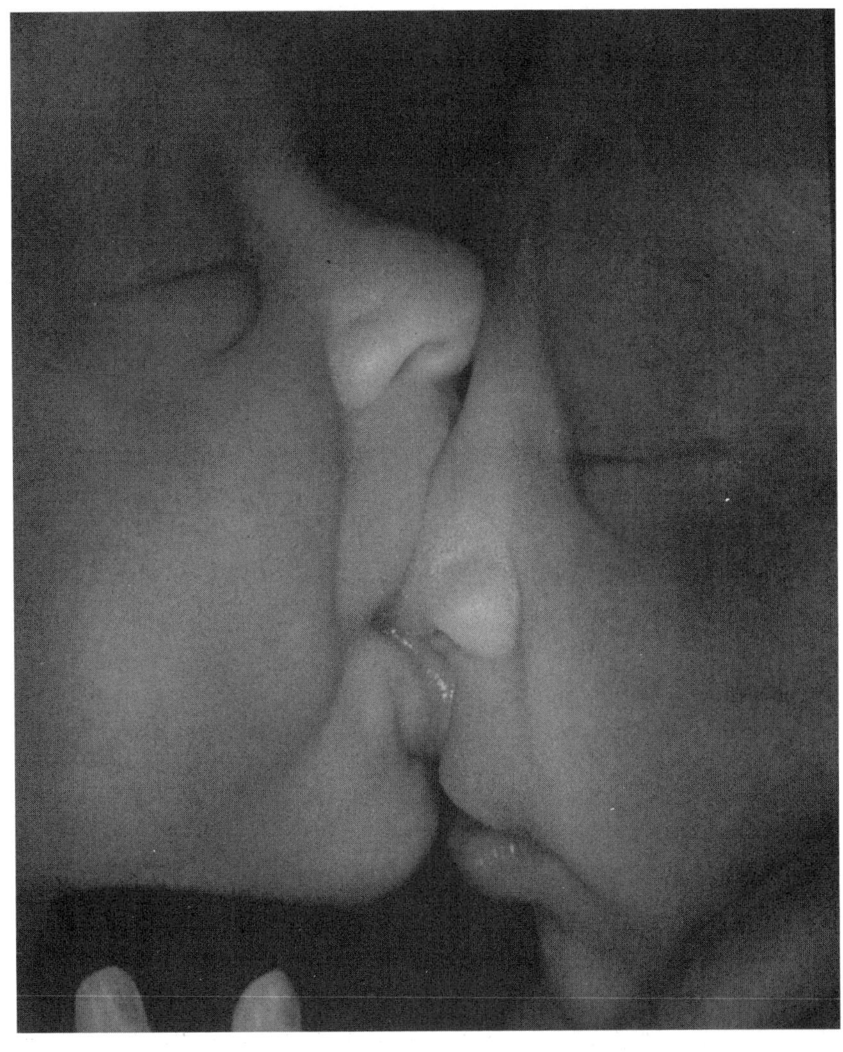

Figure 4.1 Infant identical twins one hour after birth. (Copyright by David Teplica.)

# In Vitro Babies

# 4

## In Vitro Babies

Could we produce a full-term human infant entirely in vitro, outside the body of a human mother? To examine the question of external genesis, or ectogenesis, we should first note that for many organisms, such as most fish, birds, and insects, embryonic development outside of a maternal body is the norm. In the case of most fish, eggs and spermatozoids are freely released in large amounts in the aqueous environment. Subsequent fertilization and embryonic development occur in the water as well. In birds, embryonic development takes place inside an egg containing all the required nutrients and protected by a hard shell. In this respect, mammals such as humans are more an exception than the norm. The in vitro development of a variety of organisms is very common in nature and in biological laboratories; hence is it not an unnatural notion.[1]

In vitro fertilization (IVF) is routinely performed today to help couples who are encountering difficulties in conceiving children through more traditional procedures. Plain IVF consists in presenting an egg with a sperm sample in the laboratory. An even more powerful technique, introduced by van Steirteghem, Devroey, and colleagues in Belgium in 1992, and routinely performed today, is intracytoplasmic sperm injection (ICSI). This is a technique of gamete (sperm/egg) micromanipulation in which a single sperm is captured in a microscopic glass pipette and meticulously placed (injected) into the cytoplasm of a single egg under a microscope. It is mostly done for couples in which the man has a low sperm count, as low as a few thousand spermatozoa per milliliter of sperm, as opposed to a normal level of 200 million spermatozoa per milliliter. Over 10,000 babies have been conceived by ICSI since 1992. In passing, it should be noted that once the technique for separating sperm bearing the X or Y chromosomes—the chromosomes that determine the sex of the offspring—is perfected, ICSI will allow preselection of the sex of one's offspring.

Thus we have the ability to produce the first stages of a human being in vitro, starting from a pool of male and female donor cells, up to a few days of life. At the other end, progress in medicine ensures very high rates of survival for premature infants, as young as six months' gestational age. The current Guinness record is twenty-two weeks. Producing in vitro babies is therefore a matter of closing the gap between a few days and six months or so. Such a gap is likely to continue to shrink a little through standard progress in medicine alone. To close the gap entirely, we must be able to provide an artificial developmental environment that mimics, or even surpasses, the natural environment provided by the mother: an artificial womb or amniotic sac.[2]

Under natural conditions, after fertilization the external cells of the growing embryo develop into the embryo portion of the placenta. The placenta is the vital link between embryo and mother. It is a vascular organ that unites the embryo (and then the fetus) to the maternal uterus and mediates its metabolic exchanges. By the end of the third gestational week, the embryo has a beating heart. After three months, the umbilical cord consists of three intertwined blood vessels in a fatty sheath. Two small arteries carry waste products and oxygen-poor blood from the fetus to the placenta, and a single larger vein carries nutrients and oxygen-rich blood to the fetus.

An artificial womb should have the proper mechanical properties and stimuli,[3] and should be kept at the constant temperature of roughly 37 degrees Celsius. The most difficult part is to ensure that the proper nutrients and hormones are delivered to the embryo at the proper times, and that metabolic wastes are removed accordingly, as occurs naturally through the placenta and umbilical cord. This requires bioengineering, possibly using and cultivating existing biological components, a complex communication channel equivalent to an artificial placenta-umbilical cord. In addition to oxygen and other well-understood basic nutrients, the mother provides the embryo with a complex array of proteins according to a precise and complicated schedule. Although the orchestration of these factors is still poorly understood, it is clear that one possible strategy would be to take samples from the placenta and/or umbilical chord in pregnant women at different times. There is no lack of potential sampling points because, on any day, several million women are pregnant. Whereas such sampling procedures involve some degree of risk, the technology could be improved as well as restricted to women in whom sampling is necessary for medical reasons, as in today's practice.

In an early technological stage, biological samples could be used directly to provide the basis for an artificial nutrition system mimicking the natural one. At later stages, the samples could be analyzed further and provide necessary clues for de novo synthesis of required nutrients. This would amount to a gigantic effort, but it is well within reach of current technology. Artificial synthesis of these proteins and their production in large quantities is also a formidable task, but similarly possible with simple extension of current technology. Today, we can synthesize small proteins entirely from scratch, and we can produce virtually any protein by inserting the corresponding DNA and regulatory regions in the genome of a host organism. Mass production of a given protein is certainly a costly and lengthy process, but one that has been mastered by several pharmaceutical companies. On the embryo side, we saw in chapter 3 available techniques that can provide a snapshot of all the genes which are active in a cell at a given time, as well as their product levels. In time, tools of this sort should lead to a complete picture of development and its conditions. Obviously, many of these studies would be conducted first and/or in parallel on animals.[4]

In light of all these facts, it is reasonable to think that there are no insurmountable technological obstacles to full ectogenesis. If a reasonable amount of resources were devoted to this problem, it would be surprising if a solution could not be produced within 200 years (in my opinion, in much less). But a century or a millennium makes little difference from the viewpoint adopted in this book. Furthermore, production of human babies in vitro raises somewhat less problematic issues than the manipulation and potential alteration of who we are. After all, whether in vitro or in utero, human babies are still human babies. It is even easy to imagine many situations in which this technology could help save lives, help infertile couples, and reduce the discomfort and complications associated with pregnancy.

From an ethical standpoint, the idea of raising a human embryo/fetus in vitro for nine months is not particularly shocking per se, *once good technology is available.* It does not require altering the DNA or the personality of the new human being in any major way. And again, this is already done for the first few days of life, and from the sixth to the ninth month. Problems, of course, arise with the experiments required to develop the technology. Some required experiments could be performed on animals progressively closer to humans all the way to chimpanzees and gorillas, although this is bound to raise issues, for instance, with animal activists. One argument is that once the technology is available, it could be used to help repopulate endangered species. But if the final goal were to produce human babies in vitro, sooner or later some experiments would have to be carried out on humans in the same way this was done to achieve IVF. The main considerations might be differences in stages of embryonic development that are involved, which are closer to those encountered in abortions.

A number of these experiments are noninvasive or not too disturbing, because they require studying only basic cell processes, the proteins involved in development, and so on. But at some point, experiments would have to be carried out on human embryos, for instance, after IVF. It can be expected that the initial mortality rate would be high, and that is where the main problem lies. In terms of human development stages, such deaths would be roughly equivalent to an abortion, which nowadays is typically done between the first and third months. The real difference is that in the case of abortions, fertilization generally does not occur with a premeditated goal of aborting or with the knowledge that abortion might be a likely outcome. At least at some stage of technological development, scientists would have to try to maintain an embryo alive in vitro for longer and longer periods of time, knowing that their chances of failure are still relatively high. These issues surfaced with IVF and will have to be thoroughly addressed when the time comes, probably within a few decades. Some help could come from abortions themselves, and by a strange turn of fate provide a bridge toward those who are against abortion. Because abortion is currently widespread, scientists could refine the technology in part by trying to keep aborted embryos alive and bring them to maturity. Such attempts, of course, raise other ethical, economic, and legal issues surrounding the fate of the potentially surviving baby. Such issues are well beyond our scope, and are mentioned to illustrate the complexity of the vast panorama of possibilities opening up at the dawn of the information age.

It is also possible that the first successful ectogenesis procedures could result in babies who have spent the nine months before their "birth" under suboptimal developmental conditions. It is well known, for instance, that premature babies are more likely to be afflicted by a number of problems, some of which can have lifelong consequences. If one day we decide to learn how to build developmental environments that equal or even surpass the one provided by the maternal womb, we are bound to make a few tragic mistakes along the way.

One interesting side observation has to do with one of the existing theories attempting to explain why the human brain is not larger than it is. According to this theory, the size of the woman's pelvic bones and birth canal force an upper limit on the diameter of the human skull at birth. It is obvious that producing babies in vitro does not involve vaginal delivery; hence such a constraint completely disappears.[5] It is conceivable that if this theory were true and if evolution continued to exert its pressure, humans with progressively bigger brains and perhaps greater intelligence could arise. The emergence of such beings would challenge our current notion of intelligence; however, this is not a fundamental point, because humans are on the verge of completely altering the evolutionary process. Furthermore, our notion of intelligence is about to be challenged in many other ways.

## What Happens to Sex?

*I have the greatest aversion for any form of sexual stimulation or pleasure.*

—Paul Erdös, mathematician

*Viagra is a great recreational drug.*

—Hugh Hefner, founder of *Playboy* magazine

Recently, I was describing the ideas about in vitro babies to a friend, a well-respected bioinformatician, to whom the idea did not seem to have occurred before. His first reaction, half-startled and half-amused, was to ask, "What happens, then, to sex?" What he meant to say, of course, is that from the day babies can be produced entirely in vitro, sexual intercourse viewed as a reproductive act becomes entirely unnecessary. Furthermore, shouldn't we be worried about losing what many of us would consider an enjoyable and even defining activity?

Judging from our behavior, our gossip, our media, and our art, and from the sometimes strange laws, customs, and rituals surrounding them, there is no question that sex occupies a very special place in the human psyche, behavior, and society. Ectogenesis definitely raises some issues with respect to sex as we know it. But it is only the last step in a long chain of events that started many centuries ago. Indeed, biotechnology's challenge to sexual matters is not something new. It dates back to the beginning of contraception. Condoms are perhaps one of the most widely used biotechnological breakthroughs.

In fact, contraception, abortion, artificial insemination, IVF, and in vitro babies should be viewed as milestones along the same path. The first human artificial insemination on record was performed in 1791 by a British doctor John Hunter. The first animal IVF was performed in 1878. And the first successful human technology was developed in the United Kingdom in 1977 by Steptoe and Edwards, who exposed a woman's egg to millions of her husband's sperm under the microscope. Two days later the fertilized egg was transferred into the woman's uterus; nine months later, Louise Joy Brown was born. Since then, over 300,000 humans have been conceived in vitro.

These sexual biotechnological milestones progressively give humans more freedom and control, and perhaps more balance between the sexes. On the other hand, they also progressively challenge our sexual nature. And it should come as no surprise that historically each of these steps was met with great resistance from religious organizations and other establishments. Contraception firmly established the decoupling between sexual activities and pleasure, on the one hand, and reproduction, on the other. Abortion decoupled a mother's fate from the fate of her embryo. Childbirth is no longer the inevitable consequence of becoming pregnant, nor is a mother's death in the case of a life-threatening problem related to pregnancy. But even with contraception and abortion, sexual activity remains a necessary condition for reproduction. Artificial insemination and IVF remove this necessity and also offer a solution for other infertility problems and constraints, opening the door for more flexible and free forms of genome mixing. The last step in the series, full ectogenesis, would remove the constraints imposed by pregnancy itself.[6]

Sexual and reproductive issues have long affected our societies in ways that created tensions between the sexes and were not always favorable to women. After all, even today in many countries men earn higher salaries than women for the same jobs. This is hard to justify from first principles in democratic societies, which are supposed to be founded on equality among humans. Steps in the progress toward ectogenesis have the potential in the long run for reequilibrating the balance of forces between men and women, and relieving some related gender pressures. Whereas all these steps impact both men and women to some extent, contraception, abortion, and artificial insemination can be viewed as more liberating for women, providing them with more control over their lives and reducing the disparity with respect to men. And whereas ectogenesis removes the chores of pregnancy for women, it is also perhaps as liberating for men, achieving a completely level and equal field between the sexes at the purely reproductive level. Neither gender retains a form of control or power over, or dependence on, the reproductive aspects of sex based on anatomy and physiology alone. Either member can in principle decide to "have a child" on his or her own terms and proceed with the choice, as long as there are sperm and egg donors. Even the latter condition can be relaxed with further technological advances, as described below.

To be fair, the problem of the relationship and equilibrium between the genders is exceedingly complex, and I do not mean at all to oversimplify it. One of the contributions of modern biology across species was in fact to uncover the differences—sometimes even the opposition—between the sexes and their strategies, rather than the similarities. If nothing else, the most recent trends in the area of male-female relationships are to recognize these differences and see how they can be integrated into satisfying relationships. Equality between genders may remain a valid ideal of justice, but is not necessarily a guide for practical daily matters. The biotechnological breakthroughs noted above have the potential for improving the equilibrium, but also for deteriorating it. Ectogenesis would free women from pregnancy and provide, for instance, more equal levels of employability; it could also have the potential for significantly deteriorating the social image and condition of women. After all, with in vitro babies, one of the key attributes and values associated with femininity would disappear.

As with any other technology, full ectogenesis may be very expensive at the beginning and reserved for a small number of wealthy people. But in time, with the proper technological advances and pressure to apply them, in vitro babies could conceivably become cheap and safe, possibly cheaper and safer than regular pregnancies. The liberating and equilibrating aspects, however, are to be contrasted with the troublesome questions raised above. What if sex lost its raison d'être? What if it became totally unnecessary for the reproduction of the human race?

Sex is omnipresent in our mental life and seems to play a role that goes well beyond reproduction. How much of that is a by-product of an evolutionary trick to keep us interested and ensure a good rate of reproduction for the species is highly unclear. But what is clear is that if we view sex from a rational, information-processing point of view, it appears to be an awkward and cumbersome, almost bizarre, business. Consider, for instance, two fairly large computers, over 100 pounds each, interested in merging or copy-pasting two of their respective programs. Imagine that they had to engage in some form of complex courtship. Imagine the courtship culminating in some acrobatic act by which one computer would physically mount the other, produce several billion copies of its program, and then insert its hard drive into the other computer for the transfer of these copies. Wouldn't you find that a little baroque?

From a biologist's perspective, a substantial fraction of the strangeness of sex can be attributed to a single element: water. Life has been remarkable in adapting itself to many different environments, even those characterized by extreme conditions, such as very low or very high temperatures, acidity, and so on. Life as we know it, however, cannot blossom in the complete absence of water. This is true all the way down to the molecular level. For instance, water molecules are required in the most basic energetic reactions, such as photosynthesis, the folding of proteins, and diffusion processes. This explains why during sexual reproduction the combination of genetic information from two different organisms must occur in a moist environment. This is easily achieved for fish—male and female gametes are ejected and combined externally in the surrounding water. But animals living on dry land must establish a close physical connection between a male and a female of the same species through which DNA can be transferred and processed safely in a moist environment.[7] In other words, the computers must come in close physical contact. Like mammals, birds and reptiles require a form of sexual contact to transmit genetic information. Frogs and other amphibians are an interesting intermediate case: they can live on land, but reproduce in water, and during embryonic development they have gills and swim like fish. But what about plants? How can they avoid dehydration without fornication? Indeed, plants initially seem to follow a mechanism similar to that of fish: large numbers of cells, or spores, carrying genetic information are released in the air and are often transported over long distances, typically by winds or insects. Male spores are surrounded by a protective shell. When they land on the female pistil, under the proper conditions they grow a kind of tube into the pistil to deliver their genetic material safely. Hence even plants need a form of close physical contact to reproduce.[8]

Thus, mating games and sex are necessary to establish physical contact between two individuals of the opposite sex who would otherwise remain separated, so that genetic information can be safely exchanged in the presence of water. This is a reasonable explanation of why evolution has had to find solutions that are ingenious, but also somewhat cumbersome.[9] With modern technology, genetic information can be manipulated, copied, stored, transmitted, shared, and altered without recourse to courtship or sexual intercourse, and with unprecedented degrees of speed, precision, and efficiency that are constantly being refined. Water is never a problem inside a laboratory container or syringe, and is better avoided inside a computer memory. From a functional standpoint, sex is on the verge of becoming entirely obsolete, at best an entertaining evolutionary relic that might remain fashionable or economically viable during a transitory phase of a few hundred years.

But what about orgasm? Orgasm is most likely another evolutionary trick to reinforce the human sexual experience and increase survival rates.

It is easy to imagine how technological developments could affect or replace orgasms as well. For instance, new drugs or new means of electrical or other stimulation could be developed in time that could provide humans with an alternative to sexual orgasm. I do not mean, of course, to deprive humankind of one of its most cherished activities, or to reduce sex to its reproductive aspects only and ignore other essential dimensions, such as love. I do not mean to put a number of television talk show hosts, politicians, and members of the press out of business either. But it is difficult not to wonder if sex is not an evolutionary accident that in the very long run will prove to be less important than we have been led to think during the evolutionary bootstrap phase of our history.

Many other areas of human sexuality, reproduction, and life cycle are affected by biotechnology, often in what seem to be paradoxical ways. These effects are becoming increasingly apparent, and would require a book of their own. Examples that have been prominent in the news recently range from the selection of the sex of sperms or embryos, to new treatments for impotence or infertility, to modifications of women's fertility cycles and age limits at which pregnancy or child delivery can be achieved. Two prominent cases in the news in 1990 were the first baby born to a virgin mother after artificial insemination, and the first baby born to a postmenopausal mother in Italy. And soon it should become possible for women to freeze some of their young eggs and use them for a pregnancy later in life through ICSI.

On a different but related note, in 1999 a woman in Los Angeles gave birth to a girl conceived with the sperm of her husband that was retrieved thirty hours after he died of an allergic reaction, frozen, and used fifteen months later to fertilize her eggs. Arthur Caplan, director of the Center for Bioethics at the University of Pennsylvania, indicated at the time that in at least forty-five cases in the United States, someone requested that sperm be retrieved from a dead man, but that no births had resulted. That same year a couple in London tried to obtain a vial of frozen sperm, from their dead only child, from a fertility clinic in order to have a grandchild through a surrogate mother. In addition to ethical issues, such delayed postmortem births raise many legal issues, such as the status of the child with respect to inheritance. Without ectogenesis, the symmetrical situation involving postmortem eggs from the mother necessarily implies a surrogate mother. And using both frozen eggs and frozen sperm from deceased individuals generates immediate orphans in the laboratory.

In the next chapters we will see how human cloning and manipulation of DNA further challenge our sexual identities and roles from a different direction.

Figure 5.1 Three generations of mice cloned by the Honolulu technique. The second level combines both the second and third generations, demonstrating the magnitude of the process. (Courtesy of ProBio America, Inc., and the University of Hawaii.)

# Human Cloning

# 5

*The cloning of humans is on most of the lists of things to worry about from science, along with behavior control, genetic engineering, transplanted heads, computer poetry, and the unrestrained growth of plastic flowers.*

—Lewis Thomas, molecular biologist

*The fuss about cloning is rather silly. I can't see any essential distinction between cloning and producing brothers and sisters in the time-honored way.*

—Stephen Hawking, cosmologist

## Cloning

Cloning refers to the operation by which an organism genetically identical to another one is produced. From a computer standpoint, cloning is the operation of copying a program or, more precisely, copying and executing it in the proper environment. It is important to realize that genetically identical organisms are not at all rare in nature. This is the case with cellular and subcellular organisms, such as bacteria and viruses, but also with several multicellular organisms, including fungi, some insects, and many plants for which reproduction is not sexual, or for which both sexual and asexual mechanisms of reproduction coexist. Bacteria, for instance, can reproduce both sexually and asexually. Usually, when environmental conditions are right, reproduction is asexual. When conditions become difficult, sexual reproduction kicks in, presumably to increase the level of genetic variability, and thus the chance of creating genomes that can survive, or even strive, in the harsh environment. In asexual reproduction, variability can be introduced by other mechanisms such as mutations. Even in humans, identical twins are precisely characterized by the fact that they have essentially identical DNA. And there are even natural cases of identical human triplets.

Cloning in the laboratory has a long history, briefly reviewed in appendix B. Clones of various cells and unicellular organisms have been routinely produced, selected, preserved, and studied in petri dishes. In fact, many advances in molecular biology resulted from the ability to study a number of variables while keeping the underlying genomic DNA as constant as possible in a population of clones. Even in the case of large mammals such as cows, we have known how to produce twins in the laboratory for several years. Thus genomic identity is not a stranger in the natural world or in the laboratory. And in many ways cloning a human is like producing an identical twin, but *on demand.* It is this subtle aspect related to volition, together with the possibility of producing clones on a large scale, that is troublesome, not the mere existence of isolated identical twins.

Cloning became a particularly hot topic following the announcement of the cloning of the sheep Dolly. The public announcement appeared first in the *Observer* on February 23, 1997, followed four days later by an article in *Nature.* Dolly was born on July 5, 1996, cloned by two scientists at the Roslin Institute near Edinburgh, Ian Wilmut and Keith Campbell. She was cloned using a mammary cell from a six-year-old sheep. It has been said that she was named, accordingly, after the singer Dolly Parton. The success of the operation was originally cast into doubt but was confirmed by several laboratories through direct analysis of Dolly's DNA. Since then, several groups have reported successful cloning of several other mammals, including mice, cows, and monkeys (in a weaker form, called artificial twinning, discussed in appendix B). Such announcements of course raised the issue of whether such techniques could and/or should be applied to human beings, and what the consequences would be.

From a technical standpoint, cloning involves three basic steps. The first step consists in removing the nucleus and its half-genome from a maternal egg cell. The second replaces it with the nucleus of one of the cells of the individual to be cloned, containing a complete set of chromosomes. The third step consists in jump-starting the activation of the cell division cycle of the egg with the new nucleus, to begin embryonic development. The developing embryo can then be implanted into a foster mother by means of existing technology, or developed in vitro if full-scale ectogenesis technology were available. In the case of Dolly, the acceptor and the donor cells were juxtaposed and an electric pulse was used to fuse the two cells and start cell division. But the scientists had to work with several hundred cells before scoring one success. The technology is now undergoing rapid improvement; for example, a group in Hawaii developed a technique that is conceptually simpler and easier to reproduce, and has better yields. In the Honolulu technique, a very fine needle (micropipette) takes up the donor cell nucleus and quickly injects it into an enucleated egg. The egg is then set to rest for up to six hours. To trigger cell division, the egg is put into a culture medium containing strontium, which stimulates the release of calcium from the egg's internal stores. The release of calcium is the signal that triggers cell division in fertilized eggs. The technique was used to produce over fifty cloned mice, some of which were second- or third-generation clones (figure 5.1). The cloned mice were reported to be fully functional and reproducing successfully. The mere fact that two fairly different procedures were developed for mammals within a few months is strong evidence that there must be plenty of room for improving cloning technology. Many technical hurdles and possible improvements remain, but these are well within reach of current technology. Human cloning is not a distant planet, a far horizon, but a very close, unexplored land.

Several companies are working directly on improving and commercializing animal cloning technology and its therapeutic applications. One is PPL Therapeutics, the company that was closely involved with the Roslin Institute and Dolly. In the summer of 1999, Geron bought Roslin Biomed, the commercial arm of the Roslin Institute. Scientists at ABS Global cloned a bovine by a procedure that is being promoted by a company called Infigen. Several individuals and organizations have publicly stated their intention to pursue human cloning, including Richard Seed, a scientist, and a sect, the Raelian Movement. The sect, which believes in UFOs and extraterrestrials, announced incorporation of Valiant Venture, Ltd., in the Bahamas to provide cloning services called Clonaid, Insuraclone, and Clonapet, the first two priced "as low as" $200,000 and $50,000, respectively. The Missyplicity project is a $2.3 million effort by researchers at Texas A&M University to clone Missy, an anonymous San Francisco Bay area billionaire's dog. Several dozen surrogate canine mothers have been implanted with cloned embryos. The effort has resulted in the spin-off of Genetics Saving and Clone, a Texas-based company. The company charges $1,000 to collect and cryogenically store cells from a family cat or dog, and $250,000 to clone it. Prices are expected to decrease rapidly as technology improves. The company also offers gene banking services for valuable livestock and, in the near future, for wildlife and endangered species, as well as assistance and rescue dogs.

Human cloning is achievable from a purely technological standpoint. In fact, it should be much easier to achieve than in vitro babies. With the proper resources—and leaving aside ethical or legal issues—it could be achieved within a few years, if not just a few months. The trickiest step is triggering cell division, because it involves a number of biological processes that are under active investigation but are not entirely understood. In any case, it is only a matter of making sure that the proper factors and conditions are present at the right time, and many experiments can be designed to study them. More technically challenging future extensions involve cases in which the genome to be cloned is not readily available in the form of well-packaged chromosomes in a single nucleus. This may include circumstances in which only genomic DNA fragments are available, or in which the genome to be reproduced is known as a DNA sequence in a computer and not in the physical form of a DNA molecule. These cases are addressed in later sections.

One important observation is that even if cloning is restricted to animals, it is bound to have many effects on our society. First of all, it provides a way of amplifying rare genetic events and of multiplying genomes with exceptional but rare properties that may have been obtained by chance or by painstaking and lengthy breeding experiments—for example, racehorses and other pedigreed animals. People could even immortalize a favorite pet by cloning. Although one can debate—and I will—how similar two human clones can be, intuitively it is reasonable to suspect that the clone of a dog would, after some training, be very similar to the original, at least in the eyes of its owner. Human cloning may even seem attractive to some as a way of ensuring a form of immortality. Second, cloning provides a powerful tool for scientists, particularly in research on the relative impacts of genes and the environment. It opens the door to an entirely new set of psychological, physiological, and other studies. By keeping the genome fixed, it becomes possible systematically to assess the effects of environmental variables along a number of dimensions, from physical and cognitive abilities to health and life span. This is bound to be significant for neuroscience, cognitive science, ethology, and medicine.

Ethical issues raised by the development of human cloning technology are *on the surface* less serious than those raised by ectogenesis. Experiments in animals, all the way up to our closest relatives, could continue unabated. Experiments with humans would involve failures mostly at the early stages of cell division rather than with more advanced fetuses. At this level, human cloning is not much different from IVF, mere manipulation of cellular components. At a deeper level, however, it raises far more puzzling issues, because it further modifies our views of sexual reproduction and begins to seriously challenge our notions of intelligence, identity, and life and death.

## Human Cloning Taboo

*There is nothing in the Bible that says: Thou shall clone or Thou shall not clone.*
—Ian Barbour, physicist and theologian

*This has enormous potential for good, There should be no limits on human knowledge, none whatever. To those like President Clinton who say we can't play God, I say OK, fine, you can take your side alongside Pope Paul V, who in 1616 tried to stop Galileo; they accused Galileo of trying to play God too. . . . I don't think cloning is demeaning to human nature; to attempt to limit human knowledge is demeaning. It's not legitimate to try to stop cloning. What nonsense, what utter, utter nonsense to think we can hold up our hand and just say "stop." Cloning will continue, the human mind will continue to inquire into it. Human cloning will take place and it will take place in my lifetime, and I don't fear it at all. I want to be on the side of the Galileos and those who say the human mind has no limits, rather than trying to stop something that's going to happen anyway.*
—Tom Harkin, U.S. Senator (D, Iowa)

It is instructive to observe interviews of scientists on the subject of cloning. When the question of human cloning is raised, scientists invariably adopt a conservative and prudent, almost defensive, attitude. They typically state that success in mice or cows is far from ensuring the ability to clone humans, that considerably more research is required, and that cloning raises serious ethical issues and we should proceed in this area with extreme caution. Like everybody else, I heartily agree with the latter statements. But what else is new? These are also the most dull and obvious statements one can make about human cloning. Of course, several things can be said in defense of our fellow scientists. This is a complex subject that cannot be dealt with properly in a brief television interview. Furthermore, complex issues exist that have to do with politics, research funding, patents, and the image and support of science and scientists across society. The role the press and other media play in forging this image, as well as past experiences, often lead scientists to adopt somewhat defensive and low-key attitudes. Media frenzy and the volatility of public opinion do not mix well with science.

There is, however, a second, unexpected component to scientists' attitudes. Having spoken to many scientists, I have come to the conclusion that in fact many of them have not given very much concrete thought to human cloning. This surprising fact has little to do with professional orientation and is true for most of us. For all its fascination, the subject seems to blank our minds out. It makes us feel very uncomfortable somewhere.[1] And many of us prefer not to think about it in order to avoid such discomfort, without really being aware of it. It is a little as if human cloning were in a mental blind spot. People often avoid the subject with a barrage of defensive answers, such as "Why would one want to do *that*?" or "Why would they want to clone *me*?" or the more honest "I'm not sure I'm ready for that." All these answers reveal uneasiness with the topic. This uneasiness, I believe, comes from the semiunconscious realization that cloning has the potential to challenge the notion of who we are, and that is a tall order. It is the same feeling of profound uneasiness we had to face at each stage of the decentering process—for instance, when we had to realize that our planet was not flat, or not at the center of the universe. The uneasiness that comes with the unknown and with having substantially to change one's view of the world. But cloning is here, and here to stay, and it is pointless to try to avoid it.[2] We might as well begin to stir our minds in its direction if we want to try to understand its meaning and consequences. History has shown many times that aligning our models of the world with reality is in the end a more rational and beneficial solution.

## Cloning and Sexual Reproduction

In a sense, cloning entirely removes the need for fertilization, thus removing dependence on sexual reproduction and substantially blocking natural evolution as we have known it so far. Although it is bound to be expensive initially, its cost should eventually come down. When it is available on a large scale, one could envision a population of human beings that keeps cloning itself from one generation to the next, maintaining a *completely frozen and identical genetic pool.* Of course, my point is not to recommend such a strategy or to predict that it will happen—after all, human beings do like both diversity and sex. My point is that such a thing is feasible. Moreover, genetic variety and changes could easily be produced artificially, without the need for sexual reproduction, by IVF and other DNA manipulations explored in later sections.

Many puzzling scenarios and questions based on cloning technology can be raised in the very near future. Should the cloning of a human being be allowed only after the person's death, or during his or her lifetime? In the latter case, should the two come into contact? Whereas human development may also be altered in the long term by technology, in the foreseeable future, human infants will still have to be raised by human parents to a major extent. Should an individual be allowed to raise his or her own clone? And consider a couple in a society where cloning is legal. Suppose that both individuals independently clone themselves to perpetuate their genomes. Suppose that as a couple they raise their two clones. What would happen to the clones when they reach adulthood? Could they be attracted to one another, as their identical parents were? Would that be considered a problem, a form of incest, given the fact that by upbringing they are brother and sister? After all, genetically they are as unrelated as their parents were.[3]

By definition, cloning perpetuates any peculiarities, including weaknesses, contained in the genome being cloned, such as predisposition to certain diseases. In fact, these could be amplified by cloning, because what is being cloned is an adult cell in which DNA could have mutated or been modified in other ways. Without cloning, it is difficult to know if clones will be more prone to diseases or will age faster. Finally, if cloning is used to help infertile couples, the very same reasons for infertility could be propagated. If for some genetic reason a father is infertile, it is clear that his cloned son will have a higher chance of being infertile. The same could be true of other traits, such as sexual preference, which historically have been controversial. If homosexuality is confirmed to have a genetic component, it is conceivable that a child cloned from and raised by a homosexual couple—whether male or female—may have an increased chance of also being homosexual. This is just a sample of the questions we could be faced with, and that would require a substantial revision of our laws, customs, and procedures.

## Cloning and "Intelligence"

How would cloning have anything to do with intelligence? The connection is simple and not specific to intelligence per se. It applies to any other trait or ability of an organism that is at least in part under the control of genetic factors. It may be easier for the discussion to pick a physical trait, something like physical beauty, which seems to play such a substantial role in our societies, and is clearly under substantial genetic control. Cloning the James Deans and Marilyn Monroes of this world is certainly an old human fantasy[4] with, at least superficially, some obvious advantages. But from the perspective of this book, it is better to focus on a more complex but perhaps more fundamental and far-reaching trait such as intelligence.

If we look at the distribution of human beings along any ability, we get the well-known bell-shaped curve. The curve extends quite far in both directions, although samples become rarer and rarer as one moves away from the average. Human cloning offers a simple way of sampling from these extreme distributions, making exceptional genetic events as common as desired. In principle, "geniuses" of any type could be cloned. The probability of producing an individual who is extremely gifted in some activity under natural conditions may be very low—as low, for instance, as one in 100 million. Such a rare event can be arbitrarily amplified by cloning.

Conceivably we could, for instance, produce a million Einsteins at the genetic level and experiment on different ways of raising them, both together and separately. It would be surprising if in a number of cases one did not obtain gifted and creative individuals, at least in the physical sciences. We could also interbreed geniuses, in the same way we have done for a long time with domestic animals, to increase the statistical odds.

But isn't this view of intelligence a little naive? After all, intelligence is not really well defined scientifically. It is a folk psychology word. It has served a purpose throughout our history, but it also covers a complex reality that we do not understand very well. For example, we sense that education and environment play an important role in what we call intelligence. Historical and chance factors may also help shape the life and success of a genius, not withstanding other elements related to emotions and what in recent years has been termed "emotional intelligence." Cloning Einstein a single time provides little guarantee of producing an extremely gifted individual and even less of producing a gifted biologist, if that were the goal. Whereas the precise balance of influences between nature and nurture, genes and education, on complex traits such as intelligence is far from being understood scientifically, it is clear that the genetic component plays a nontrivial role. There is no question, for instance, that numerous genes control to a certain degree the anatomy and development of the brain, the density and types of cells and receptors in different areas, the neural connectivity, and so forth. The existence of a genetic component, no matter how small, is all that is necessary if the goal is progressively to amplify genetically rare events.

An example of a study of genetic factors involved in what we may call intelligence can be found in an article in the journal *Nature* (September 2, 1999) by Y. P. Tang and colleagues at Princeton University. One important protein in the brain that seems to play a major role in learning and memory is the NMDA (*N*-methyl-D-aspartate) receptor. The NMDA receptor is associated with how synapses—the connections between brain cells—change their strength as a result of synchronized activation. The change in strength is believed to be one of the physical traces associated with memory and learning. The authors created a new strain of transgenic mice, dubbed Doogie after the lead character in the TV show Doogie Howser, M.D., with extra copies of the gene that codes for one type of

NMDA receptor. In this way and through other manipulations, they induced overexpression of the NMDA receptor in the forebrain of the mice, even as the mice were aging. A battery of standard behavioral tests showed that the transgenic mice were able to learn faster and better, and to remember things up to five times longer than control mice. For instance, the scientists tested spatial learning by putting the mice in a pool of water containing a hidden platform allowing them to climb out of the water. The Doogie mice found the platform after three sessions, whereas the normal mice took six to get it right. In addition to shedding further light on the role of the NMDA receptor, the authors concluded that "genetic enhancement of mental and cognitive attributes such as intelligence and memory in mammals is feasible." One hope, of course, is to develop such ideas into human gene therapies for disorders such as dementia. Yet, as unethical as it may sound, it is difficult not to wonder what would happen in Doogie transgenic humans.

Returning to our main topic, identical twins are one of the major tools for psychological studies attempting to tease out the roles of nature and nurture in humans. Such studies suffered from the fact that identical twins are relatively rare. Cloning, of course, could enable analysis of these difficult questions on an unprecedented scale, and contribute to our understanding of how to create optimal nurture conditions, thus further increasing the chances of producing extremely gifted organisms. This is shocking for humans, but it has been a long-standing goal in farm animals, for which the practice is well under way. Consider, for instance, racehorses. Is there much doubt that the clone of the best thoroughbred, raised with the optimal diet and training, would have a very high likelihood of yielding a champion? And one could breed optimal clones together in an attempt to further amplify certain characteristics on a new scale. On a more humorous side, it must be observed that cloning poses a serious threat to the "ownership" of genomes. It is much easier and cheaper to steal a cell from a stallion than to steal or buy the stallion itself.

We all find "farming" humans a repulsive idea. And judging from our history, one can wonder if a society trying to farm humans would not inhibit the expression of the highest forms of human intelligence. But it is a simple technique for producing genetically gifted individuals and making rare genetic events very common. And ask yourself whether you would like to have a clone of yourself—an identical twin, identically raised with you—work with you on some difficult task. Have you ever wished to be able to duplicate yourself for some task?[5] Furthermore, conditions could be found in which clones could cooperate efficiently, further challenging the boundaries of their particular abilities by some form of parallel processing. Although history shows that putting smart people together does not always produce good results, the situation here is different because of the genetic identity of clones and our increasing technological ability to manipulate our environments. In addition, adult identical twins, whether raised separately or together, do seem particularly to enjoy each other's company.

Many examples of extreme cooperation of this sort are described in the literature on identical twins. Under unusual conditions, for instance, identical twins raised together have a tendency to develop their own language.[6] One of the most extreme examples is the case of famous identical twins reported by Oliver Sacks in *The Man Who Mistook His Wife for a Hat.* This case affects our ideas of intelligence, and of how the brain works, as well as their connection to genetic and environmental factors, cloning, and parallel processing. As reported by Sacks, John and Michael were famous in 1966 at the age of twenty-six. By many measures, they were mentally retarded, with an IQ of 60 and the inability to do arithmetic or even comprehend the concept of multiplication, as well as other obvious signs of emotional and intellectual retardation. They had been variously diagnosed as autistic, psychotic, and severely retarded:

*They are, indeed, unprepossessing at first encounter—a sort of grotesque Tweedledum and Tweedledee, indistinguishable, mirror images, identical in face, in body movements, in personality, in mind, identical too in their stigmata of brain and tissue damage. They are undersized, with disturbing disproportions in head and hands, high-arched palates, high-arched feet, monotonous squeaky voices, a variety of peculiar tics and mannerisms, and a very high, degenerative myopia, requiring glasses so thick that their eyes seem distorted, giving them the appearance of absurd little professors, peering and pointing, with a misplaced, obsessed, and absurd concentration.*

At the same time, however, the twins are idiots savants displaying an incongruous array of extremely unusual abilities, typically performed in emotionless ways. For instance, they exhibit a sort of documentary memory of the smallest visual details of their experience, and are capable of reporting the weather or any other event of any day in their lives. They have an unusual memory for digits, being able to memorize 300-digit numbers with ease. They also can tell on which day of the week a day in the distant past or future would fall. This is achieved apparently by the use of an unconscious calendrical algorithm that has a peculiar "visual" quality to it, as indicated by their verbal responses; the fact that their eyes move and fixate during these calculations "as if they were unrolling, or scrutinizing, an inner landscape, a mental calendar"; and the overall impression of "seeing" the answer rather than calculating it. The twins also are able to factor large numbers and recognize large primes, of which they seem particularly fond.[7]

What is also remarkable for our perspective is the fact that these twins are indistinguishable, without individuality, as if they had almost the same self and the distance between their brains were very small. If a mechanism is at work that tends to differentiate identical twins raised together, it seems not to have worked in the case of John and Michael. Because of this similarity, they seem to be able to cooperate very tightly on the tasks they engage in, and to understand one another at a deep level inaccessible to others. Their talents and cooperation also have social and emotional components. After being separated in an attempt to render them more independent and socially acceptable, "they seem to have lost their strange numerical power, and with this the chief joy and sense of their lives."

In any case, the twins exemplify many themes in this book. They provide us with a glimpse of a vast universe of extreme cases that has remained hidden and undersampled during our tip-of-the-iceberg human history. A universe of extreme cases characterized by peculiar brains endowed with peculiar computational powers. A universe where brains and selves arbitrarily close to each other can exist. A universe where brains can be tightly associated and process information in parallel. They represent a phenomenon that is very rare under natural conditions, but that could be amplified through cloning and other biotechnologies if we wanted to. In fact, one of the points of this book is that what seems to us a strange and rare computing universe is in fact the pervasive norm. The undersampled computing universe we are now accustomed to is rather the exception, a consequence of the evolutionary and historical bootstrap process by which we have come to be.

In summary, the chance of having twin Einsteins by natural means is essentially zero and has never happened throughout human history. The chance of having identical twins in the general population is very roughly one in a thousand.[8] If we put the chance of having an Einstein at 1 in 100 million, the chance of having both is 1 in 100 billion, a number larger than the total number of human beings who have ever lived![9] Under natural conditions such an event is so unlikely that it can be considered impossible. But it is trivial to replicate with cloning techniques. Thus cloning is a simple way of creating genetically smart brains and putting them together in ways that far exceed anything we have ever seen. In the coming chapters we will go one step farther by manipulating and mixing corresponding genomes.

## Cloning and Life and Death

Cloning also questions our notions of life and death. At a minimum, it provides a form of genetic immortality. Such immortality is not uncommon in nature, with organisms that can reproduce asexually, and in particular with plants such as strawberries and aloe. One can easily cut a piece from an aloe plant, stick it in the ground, and watch it grow within a matter of weeks. Thus a plant can be surrounded by genetically identical plants. With cloning technology, we could have an identical twin alive at all times. In fact, in principle we could have an entire population of identical twins alive at all times—imagine 1 billion of them, containing various realizations of our own genomes in different forms, some of which are arbitrarily close to who we are. It is this statistical population that becomes immortal, rather than the single individual.[10] The single individual is just a sample that might be hard to reproduce perfectly, but can be approximated with arbitrary closeness with respect to any reasonable measure. These approximations are treated in the next chapters. For some of us it may be reassuring to think that an identical twin will always be there, and possibly many of them. It used to be that we could pass only 50% of our genes intact to the next generation through any one child. And by the standard reproductive process, our genome would be diluted rapidly, exponentially fast, over a few generations. With sequencing and computers, our genome can be stored forever in a computer file; with cell cultures, it can be kept in a format ready for cloning at any time; and with cloning, it can be executed at any time and impersonated in any one of a continuum of possible identical twins.

Molecular biotechnology and medicine will affect our quality of life and our life spans in many more direct and pedestrian ways. We do not know exactly what the mechanisms of aging are at the molecular level. Current theories involve phenomena ranging from cumulative oxidative stress and damage by free radicals, to the progressive shortening of telomeric DNA (the DNA at the end of the chromosomes) after each cell division.[11] Under optimal "natural" conditions, the human life span is believed to be in the range of 120 years. This estimate was derived from a number of considerations, including statistical ones, and the observation of the age reached by some of the oldest humans. With better nutrition, living conditions, and medicine, the life span has been increasing—by thirty years over the last century—and will probably continue to do so toward and beyond this upper limit, provided environmental conditions do not deteriorate too drastically in unforeseen ways. But the question of interest from the perspective of this book is not whether we can extend our life span by a meager century or so.[12] Or whether we can then further increase it by another century. The question is whether this limit can be pushed much farther, all the way to what could be called immortality.

Indeed, we have many possibilities for increasing life span well beyond 120 years through "nonnatural ways." One approach is a calorie-restricted diet, which worked on all the animals it was tried on. In small rodents, life span was increased by up to 40%. Results in monkeys seem to go in the same direction, although it is premature to draw firm conclusions. Rhesus monkeys can live up to forty years in the laboratory, and experiments with them were started in the 1970s.

On a microscopic scale, methods of extending life include gene therapy in its broadest sense, as well as methods to stimulate growth and repair mechanisms, and to slow or even reverse aging. A gene was found in the fruit fly that, when mutated, increases the fly's life by over one-third. This is the "Methuselah" gene discovered by Seymour Benzer's group at Caltech. Methuselah, a Hebrew patriarch mentioned in the Bible, is said to have lived to the ripe old age of 969. *Drosophila* flies normally live sixty to eighty days. Flies with a mutation in the Methuselah gene live over 100 days. In correlation with longer life, the gene also confers better resistance to various forms of stress, such as food deprivation, excessive heat, and exposure to oxidative damage. A similar gene—with a more boring name (daf-2)—was found in the roundworm *C. elegans*. Because fly and worm genes often have human homologues, it is likely that genes controlling the life span of vertebrates also exist. Of interest, a link between sensory perception and life span has been found in the worm. When genes involved in smell and taste are altered so that the worms are unable to receive environmental signals, the worms live up to 50% longer (Apfeld and Kenyon 1999). Thus life span might also be regulated by environmental cues.

On a macroscopic scale, artificial and artificially grown organs, tissues, and organ transplants could be used to address aging or damage of particular body parts. We can grow skin and cartilage tissues to treat people with severe burns or damaged cartilage. We can also grow bone. Growing complex organs such as a heart or a kidney, which comprise several tissues, is much more difficult. And at the molecular level, we are just beginning to understand how genes control organ shapes (e.g., Blelloch and Kimble 1999). The first successful cardiogenesis experiments were carried out in amphibians (Ariieumi et al. 1996). Similar experiments were successfully performed on bladders and arteries (Ferber 1999a; Niklason et al. 1999). Artificial cardiac tissue has been bioengineered by researchers at the Massachusetts Institute of Technology, using rat neonatal cells and a simulated space environment based on a bioreactor that rotates constantly to keep the cells in a sort of endless, gentle free-fall state (Carrier et al. 1999). The tissue resembles cardiac tissue and has the proper electrical properties that allow it to contract. It could eventually be used to repair damaged heart tissue inside the body, test new drugs, and study cardiac tissue development and function. The technology must be improved in several directions—for instance, in terms of tissue size and inclusion of blood vessels—and extended to humans, but it provides a sense of the possibilities.

One of the greatest promises in this general area is the use of embryonic stem cells, first obtained in humans by Thomson and coworkers in 1998, and reviewed in some detail in appendix C. Stem cell research was voted "breakthrough of the year" by the editors of *Science* at the end of 1999. These are cells derived in general from an early embryo that are totipotent (i.e., capable of developing into virtually any cell of an organism under the proper conditions) (see also Solter and Gearhart 1999). The advantage of stem cells over more artificially designed systems is that they have the built-in hardware and software required to develop an entire tissue, organ, or even organism. In time, we could learn how to induce the differentiation of embryonic stem cells and how to build artificial organs and tissues from scratch. Needless to say, this would have a profound impact on medicine and on how to prolong life. This line of work is rapidly progressing in animal models. Researchers have also found that stem cells taken from adult tissues can be converted into different cells. In rats, for instance, bone marrow stem cells can differentiate into liver cells (Petersen et al. 1999), raising the possibility of one day using a patient's own marrow to repair his failing liver. If embryonic stem cells from a given individual were not available through cloning or some other procedure, we could also imagine modifying the genomic DNA of a human embryonic stem cell to ensure that any developed tissue or organ be compatible with the acceptor.

Transplantation of many organs, including hearts, lungs, livers, and kidneys, is routinely performed for medical reasons. Some of these procedures, including kidney and (partial) liver transplants, have even been performed with live donors. Severed limbs and other appendages have been successfully reattached. Even hands with all their complexity have been successfully transplanted, and in January 2000, an international team of eighteen surgeons performed the first double arm transplant. Several new transplants, such as testicular transplants, are in the works. A number of organ transplants from animal to human (xenotransplants) have been attempted since the early 1960s.[13] Today, pig valves are routinely transplanted into human hearts. Artificial arteries can be created with biological material taken from pigs and cows. Cow neck valves have been used in the hearts of infants, and liver cells as well as entire livers from genetically modified pigs have been used in xenoperfusions to filter blood from a patient with liver disease in a process similar to kidney dialysis.

Overall, pigs are good candidates because many of their organs are similar to their human equivalents. Transgenic pigs, those with altered

DNA possibly containing human genes, could be engineered to avoid rejection. Cloning could be useful in the industrial production of such engineered pigs. In addition to biological and biologically engineered organs, artificial organs, such as artificial hearts, and artificial blood, bones, and skin are beginning to appear. Artificial chip implants are being tried in blind people and paraplegics. Note, however, how even simple organ transplants begin to challenge our notions of identity and humanity. This is not obvious when we think of single-organ transplants, which are quite common. But in principle, the technology is here so that limbs, liver, kidney, heart, and lungs can be transplanted into a single individual. When in addition some of the organs are of animal or artificial origin, the challenge becomes striking.

All the technologies discussed so far could benefit from, or be combined with, cloning. As described for the case of pigs, cloning could be used to produce expensive genetically engineered organisms. By providing genetically invariant populations, cloning is an ideal tool in medicine for studying the impact of anything—diet, lifestyle, environment—on our health and life span, as it is in psychological studies that try to tease out the effects of nature and culture. It would be important in developmental studies and could be used to provide fresh and compatible biological material, from cells to tissues and entire organs, in a number of therapeutic approaches and organ transplants.[14] Another possibility that raises many ethical issues would be creating brainless clones to be used only for their organs. Brainless frogs and mice have already been obtained.

Needless to say, farming clones merely as fresh organ donors is totally appalling and against all our ethical values. But the reality is that the technology for producing clones, and hence an arbitrary supply of perfectly matching donor organs, is here. In time, we should be able progressively to achieve the same results in more ethical ways, using a number of possible approaches. We might start using cells taken only from the very early developmental stages of a clone—when many believe a conscious human being has not yet formed and abortion is legal. We might learn how to grow isolated organs in vitro, using, for instance, embryonic stem cells. We might learn how to bioengineer better animal donors. We might learn how to bioengineer organs using various combinations of biological and nonbiological substrates and technologies, further blurring current categories.

Finally, one common misconception about cloning and its relation to time and age is ingrained in the old way of thinking about these issues. People often say things like "Yes, you might be able to clone me, but after nine months you will get a genetic copy that is a baby rather than an adult of my age." This is true if we start the cloning operation now, and if technology to speed up development is not available. Although such technology might become available one day, the difference in age is only a minor secondary effect due to a transitory phase. In other words, if the cloning request is given sufficiently in advance, it is easy to meet any age requirements. For instance, if in 2080 we want to have a thirty-year-old clone of Einstein, all we have to do is to start a cloning operation of Einstein's genome around 2050. In time, we could imagine a gigantic cloning "machine" that continuously outputs in parallel new, young clones of all possible genomes—an extreme form of family planning! At equilibrium, such a process would yield at any time an arbitrary supply of clones of any age for any given genome. Thus, if desirable, any living individual would be assured of having contemporary clones of himself of exactly the same age, and in fact of all possible ages. Such an extreme machine may seem impractical, but it would not be too hard to have a slightly weaker approximation based on a repository of genomes, in either computer or stem cell-culture format, where genomes can be selected in advance and executed at the right time ad libitum.

In any case, we can progressively approximate some form of immortality, at least for most of our bodies. This could take several hundred years. But what about our brains? After all, that is the center of our concept of self. Could we have brain or self transplants? Whether brains can also become immortal, and in what sense, is addressed in chapter 9.

## Cloning and Identity

In the same way that it questions our notion of life and death, cloning questions our notions of identity. It is worthwhile to pause here and ask ourselves whether and when we ought to say that two organisms, and in particular two human beings, are identical. At one extreme end of the spectrum we could require identity at the level of single atoms; that is, each atom of the first organism should match a corresponding atom in the second organism. This of course seems excessive—we would not be the same person after a good meal, a haircut, or surgery. Furthermore, such a condition could not be maintained in time because the two atomic assemblies would have different spatial coordinates and therefore would immediately begin to diverge. At the molecular level we are in a constant state of random fluctuations and changes. All molecules in our body are said to be replaced every seven years or so. Thus, to a certain extent our sense of self is robust with respect to molecular fluctuations.

At the other, weaker end of the spectrum, we have genetic identity when two organisms have exactly the same genome. This is the case, for instance, of identical twins. Identical twins are extremely similar on all possible scales, from macroscopic to microscopic.[15] However, some variations are bound to arise from a variety of sources, such as small differences in the boundary conditions during embryonic development, viral infections, and, more generally, life histories. Identical twins, for instance, do not have identical fingerprints, although the patterns of ridges on the skin of their fingers are correlated. Similarly, they generally do not have the same exact height, although the measured height correlation is 0.94. (This number shows that the heights are in general very similar but not identical—the correlation would be equal to 1 if they were identical. The same correlation measured on randomly selected individuals would be zero.)

Clones are like identical twins. In fact, technically speaking, and contrary to popular belief, they might be a tad less similar. This is because some differences between the DNA sequences of two different cells may be present in the same adult organism as a result of random events such as mutations and mistakes during replication. Thus the genome of a clone can depend very slightly on the specific cell used for cloning. In most cases, this so-called somatic polymorphism is very small, but the number of differences increases with the level of differentiation of the cells, and can be far from trivial in some special cases. In addition, the differences in the starting conditions for the embryonic development of an individual and his clone can be more pronounced. The egg is a very large cell that contains all the necessary elements for processing genetic information and bootstrapping the developmental process. In other words, not all eggs are identical even if the chromosomes they harbor are. Thus clones produced from eggs of different donors, or from the same donor but taken at different times, may show a higher degree of variability than clones produced using eggs taken from a single donor at the same time. Identical twins initially share the same egg with, for instance, the same mitochondrial DNA.[16] Differences of this nature have been observed in cows, where markings on the hide of a clone can differ substantially in position and size from those of the original animal. These differences are likely to result from fluctuations in the developmental morphogenetic processes rather than in genomic DNA.

But by far the main differences between an individual and a clone are in their brains. Even if raised together, two clones end up having slightly different experiences and memories, and therefore different selves. For many of us, two clones are two different persons, and therefore we have little to fear for our cherished notion of self. But is it really that simple? Studies of identical twins have shown over and over a degree of similarity that goes beyond the physical, and includes many traits, habits, and behaviors. And think how you would feel if you were sitting in a room with ten clones of yourself! Would it not be a little unsettling? And what about living in a town with 1 million clones of yourself? Such events have essentially zero probability of occurring under natural conditions, but become a simple matter of resources once the technology is available. The point is not to build such worlds à la Huxley,[17] but to note that they are possible. As we will see in the next two chapters, we can also imagine many ways of influencing brain variations. Such a large population of human clones corresponds first to a single point in genome space: all its members have identical genomes by definition or by construction. On the other hand, for notions such as personality and self, the populations should be viewed as a cloud of points. The cloud can be made as dense as desirable by more sampling and more experimentation. Although each clone in the cloud is a separate and in some sense irreproducible individual, individuals can be produced that are as close as possible to him with respect to whichever reasonable metric one wishes to use. Therefore the picture that begins to emerge with simple cloning, and is expanded in the next chapters, is that who we are belongs more to a continuum than to a discrete space. During the evolutionary bootstrap we have been led to think in terms of a discrete space, in terms of well-defined, isolated individuals, because only discrete and relatively distant samples were available in a natural setting. But this is not the reality.

The reality is that arbitrarily close human beings can be produced, and that for any human being we can construct clones that are arbitrarily close to him or her. The underlying continuity was there all along but has remained hidden from us until today, creating an illusion—the evolutionary illusion. Up to now, our brains have evolved in this paradoxically artificial—or, rather, undersampled and misleading—universe that was a by-product of the sampling bias introduced by sexual reproduction and genome shuffling.

Figure 6.1 Eng and Chang, the original Siamese twins. (Courtesy of North Carolina Collection, University of North Carolina Library at Chapel Hill.)

# Siamese Twins

# 6

*The sympathy existing between these two brothers is so close and so refined that the feelings, the impulses, the emotions of the one are instantly experienced by the other. When one is sick the other is sick; when one feels pain the other feels it; when one is angered the other's temper takes fire.*

—Mark Twain, writing about Eng and Chang

"Siamese twins" is the name popularly applied to twins who are physically conjoined in a manner not incompatible with life or activity. Under natural conditions, it is the result of an imperfect splitting of a fertilized egg that would have resulted in identical twins.[1] The name is derived from famous twins, Eng and Chang Bunker, born in 1811 in Siam (now Thailand) (figure 6.1). The men were joined at the sternum by a thick, ligament-like appendage. They barely escaped death by euthanasia. When Eng and Chang were teenagers, they began traveling with two agents, giving lectures and demonstrations all over the world, especially in the United States. When they decided to retire from traveling, they chose Surry County, North Carolina, where they became successful farmers. In 1843 they married two daughters of a neighboring family, Sarah and Adelaide Yates. Eng and Sarah had six sons and five daughters. Chang and Adelaide had three sons and seven daughters. The men died in 1874, at the age of sixty-three. Chang preceded Eng in death by about two and a half hours. An autopsy showed that Chang died of a blood clot in the brain, and although it was never officially confirmed, many indications are that Eng died of shock. A grandson of Eng became president of the Union Pacific Railroad, and a grandson of Chang became an air force general.

It is quite interesting to see how the topic of Siamese twins has been treated and represented in strange ways throughout history and in the media.[2] Even today these individuals are treated with a kind of fascination reminiscent of the fascination with monsters, witches, and satanic rituals and punishments, combined with a Guinness Book of Records carnival atmosphere. One side of the fascination is obvious—Siamese twins challenge our own boundaries and our definition of what is human. But it is high time to get rid of our defensive reactions based on an old "magic" view of the world and take a closer scientific look at the reality.

The first thing to understand is that Siamese twins, under natural conditions, are the result of a developmental accident by which two embryos become fused. This typically happens with identical twins, which also explains why Siamese twins are of the same sex. A wide variety of cases of Siamese twins have been recorded in the literature, depending, for instance, on how the bodies are fused, which organs are duplicated, and which are shared. These parameters have obvious important consequences for the life of the twins, and for any surgical attempts to separate them.

In this respect, Eng and Chang constituted a relatively mild case of fusion. Each man had two arms and two legs, and fusion occurred only at the level of the sternum. In more extreme cases, Siamese twins can be fused at the level of the head, or have only one leg or one arm each. There are also documented cases of conjoined twins in a triplet pregnancy. With modern medicine, surgical separation is often possible, although it can raise complex ethical issues. For example, in some instances the anatomical complications are such that the lives of both twins are in danger and surgery can save the life of only one. Although surgery is invaluable in many cases, one can wonder if occasionally it was not tainted with a trace of the desire to normalize and suppress what is different and disturbing. Some adult Siamese twins have chosen not to be separated.

It is clear that Siamese twins could be produced artificially with current technology by fusing embryos, or transplanting cells from an embryo to another, in different ways and at different stages of relatively early development. This has been done in frogs, for instance. One might also be able to fuse triplets or twins of opposite sex. It is not unthinkable that methods could be developed to fuse embryos that are genetically more and more distant, although many complex compatibility problems would have to be resolved. How far could we go in producing chimerical beings? Although scientists are still far from understanding developmental programs in detail, they have already experimented with a wide range of chimeric beings and chimeric transfers.

Animal genes have been expressed in plants, and human genes in bacteria, yeast, mice, and cows. Even an entire human chromosome was inserted in a transgenic mouse. Interspecies organ transplants, such as from pigs to humans, are not uncommon. At a different level, Dolly was obtained by transferring the nucleus from the cell of a Finn Dorset sheep into the enucleated oocyte of a Scottish Blackface sheep. Cloning of a sheep using a cow egg has been tried, but with no success so far. In time, transfer of genetic and biological material across biological systems and species will become increasingly easy from a purely technical standpoint, opening unprecedented avenues for tinkering with and creating new systems. Another way of producing Siamese twins artificially is by manipulating the expression of particular genes during development. This has been done in the laboratory on frogs, where the overexpression of a particular gene has been shown to lead to frogs with two heads. Thus there is little question that we can produce a variety of beings which challenge the old taxonomic and anatomical boundaries and, in particular, which can have several brains in one physical body.

More fundamentally, perhaps, chimeric experiments have been carried out with totipotent embryonic stem cells. These are cells taken from an early embryo (blastula stage) that can be cultivated and can develop into virtually any cell of the body. Embryonic stem cells were first obtained for mice, but recently also for humans (Thomson et al. 1998). Mice embryonic stem cells are widely used in research.[3] When stem cells from a black mouse are injected into the embryo of a white mouse at the blastula stage, a black and white mouse results. The stem cells contribute to the development of the entire organism. Similar experiments have been carried out by injecting human cancer and other cells into mouse embryos.

To further this analysis, it is natural to wonder about the way Siamese twins feel and perceive themselves. To what degree are their brains different? To what degree do they have distinct selves? And to what degree do they process information together? After all, they are identical twins: they have the same genome and of necessity are raised in similar conditions, live in the same house, sleep in the same bed. Their circulatory systems often intersect, so they share the same blood, the same nutrients, the same hormones, and so forth. As a result, they must be fairly synchronized in many of their rhythms, from hunger, to wake-sleep, to puberty, and to death. Reports about Chang and Eng state that one of them used to drink a little too much and thus often got the other drunk, too. For the same reasons, and although this would lead to an intriguing literary subject, it seems unlikely that natural Siamese twins could exist with different sexual preferences. By necessity, it is also clear that the brains of these individuals must often cooperate very closely, in particular for all things related to motion in space. For instance, take the case of Siamese twin sisters, each with one leg, who are capable of walking without any major difficulty. This obviously requires precise coordination of motor commands sent by the two brains to the two limbs. And what about Siamese twins controlling one hand each? Could they ever become a very skillful surgeon, performing separation surgery on other Siamese twins?

In reality, the main study assessing psychological differences between Siamese twins reports that in general the conjoined individuals appear to be quite different. This is a surprising result because in a sense Siamese twins appear to be the highest possible degree of identical twins raised in similar conditions. At least three elements provide an explanation for this phenomenon. First, the result is consistent with the more general observation that identical twins raised in different conditions tend to be more similar than identical twins raised in similar conditions. This could be the result of psychological mechanisms that would increase the propensity for those raised together to differentiate from one another to establish their individuality. The same mechanisms could operate for Siamese twins.

In fact, and this is the second factor, a subtle methodological problem could be related to how differences between Siamese twins are assessed. It is conceivable that when they are tested, the two individuals tend to exaggerate their differences to establish their individualities. By definition, it is difficult to test Siamese twins in complete independence. It is even difficult to anesthetize one of them without anesthetizing the other. Even if great care is taken to prevent direct communication or sharing of information, it is always possible that some communication may take place "internally." And this communication could form the basis for polarization of the answers.

But the third observation is that Siamese twins, as a result of their fusion, are in fact raised in very different conditions, and in a sense in conditions that differ in a constant way. Even during early embryonic development, fusion may result in different sets of constraints being imposed on each brain at the physiological and anatomical levels. Siamese twins are never symmetric or perfect mirror images of one another.[4] In particular, the inputs and outputs of their brains, with respect to both their external and internal worlds, are radically different. For instance, the set of organs and muscles each twin controls is different. One twin may command the right arm and the other the left arm. One twin may always sit to the right of the other. Their visual systems see the world from different angles, with one head always on the same side of the other one. In humans and monkeys, it is well known that suppressing input to the brain from one eye or one finger during development leads to substantial reorganization of the corresponding primary sensory cortical areas. Thus, in reality the brains of Siamese twins are under very different conditions during development. I suspect that such radical and systematic differences and distortions are also likely to result in substantial overall brain differences.

A puzzling question is whether Siamese twins could share the control of an organ. In a weak sense, we have seen that they can share control of the composition of their blood. In general, and under "natural" developmental conditions, nerve connections from a given organ will be established with only one of the two brains, so that we cannot expect to have shared control. But what about artificial developmental conditions?[5] In animal experiments, for instance, the growing optic nerve has been rerouted toward auditory areas. It is not entirely inconceivable that in time one could create Siamese twins and reroute certain nerves from one brain to the other. Early development is a blind process without a notion of self. Furthermore, there is the case of Siamese twins who are joined at the heads. To what degree could their brain be fused, and what would the implications be? And could we one day fuse embryos associated with different genomes?

What would happen if we combined cloning and Siamese technologies? Someone, including you, could take some of your cells and produce several clones. Early in development, the embryos could be fused, creating Siamese twin or triplet clones. How would you feel if you were surrounded by thousands of single, Siamese twin, and Siamese triplet clones of yourself?

This is not science fiction—this is extremely close to being technically feasible today.

Here and everywhere else in the book, the point is not to suggest that we carry out such activities. The point is that nature and evolution have so far produced only an extremely small sample of all possible genomes and forms of life on Earth. And our brains have evolved and created a model of the world based on this limited environment. The universe of all possible living beings, however, is much larger. By manipulating DNA and development, the blueprint by which living beings are built and its execution, we can suddenly sample this vast universe of beings, only a small fraction of which has ever roamed our planet. When we sample it systematically, even only as a thought experiment, we discover many forms of life that challenge many aspects of who we are and what we do. We even find beings that have the potential for trespassing the boundaries of our mental life, toward new, uncharted directions.

Figure 7.1 DNA morphing from lion to tiger. A liger resulting from the crossing of a male LIon with a female tiGER. (Courtesy of Beverly Setlowe, Los Angeles Wild Animal Way Station.) Ligers are not fertile and tend to be somewhat nervous and uneasy animals. Tigrons—the cross of a male tiger with a lioness—are also possible, although somewhat rarer.

# Manipulating DNA

7

## The Extended Jurassic Park

*For a mammal like man, there was something indescribably alien about the way reptiles hunted their prey. No wonder men hated reptiles. The stillness, the coldness, the* pace *was all wrong. To be among alligators or other large reptiles was to be reminded of a different kind of life, a different kind of world, now vanished from the earth.*

—Michael Crichton, in *Jurassic Park*

In vitro babies, human cloning, and Siamese twins already go a long way toward shaking the foundations of our view of the world. But in many ways they are just the beginning, because so far we have given little consideration to the possibility of manipulating or altering DNA in a direct way. We have also exclusively examined intact genomes and cloning with living cells from the same species, at both the donor and the acceptor levels. Simply relaxing these constraints and expanding the cloning boundaries can shake the old foundations to the breaking point.

We have seen that we are very close technologically to being capable of cloning any *living* human being, and thus his or her genomic identity can be rendered more or less immortal. This, however, introduces a kind of asymmetry with respect to human beings who are not alive anymore. Somehow this asymmetry does not seem to be justified. How could a general concept such as genomic immortality really be dependent on the historic period at which one lived?

Consider, for instance, Albert Einstein, chosen as "person of the century" by *Time* magazine at the end of December 1999 (he died in 1955). Is it reasonable to think that just because he missed the cloning age by half a century or so, his genome could never achieve immortality? This is very doubtful. Consider the well-known fact that Einstein donated his organs to science upon his death and that, for instance, his brain has been preserved. His genome—in fact, his actual DNA—could be recovered and potentially used in a cloning operation. This of course would require finding ways of working with DNA and surrounding material in conditions that are not identical to those encountered in the living cell. Death, as well as the application of preserving substances such as formaldehyde, is bound to induce some degree of change in biological tissues. Luckily, the DNA double helix is a relatively stable molecule. Furthermore, with more advanced technologies described below, original DNA molecules are not necessary for cloning, only the corresponding sequences, and these are becoming easy to obtain, even with today's technology.

Similar techniques could be applied to DNA in other states, such as the DNA of organisms preserved in a frozen state or in amber. As the technology improves, one could begin to work with more and more ancient DNA, and also with fragments—for example, when the entire genome is available not in one single entity but as a set of overlapping fragments. These are just different ways in which the same information can be packaged. DNA was recovered from Oetzi, a man whose 5,000-year-old frozen body was found in 1991 in the Italian Alps. Egyptian mummies contain DNA several thousand years old. One of oldest known fragments of human DNA is 8,000 years old and was found in an acidic bog in Florida that had been used as a burial place. Even older DNA was recovered several years ago from a member of an extinct species: a deep-frozen mammoth found in Siberia that is estimated to be 40,000 years old. Mitochodrial DNA also has been recovered from a Neanderthal specimen. Pieces of amber have been found with trapped insects that are over 100 million years old. The oldest DNA fragment retrieved up to now belongs to a 125-million-year-old beetle.

Ancient DNA fragments raise several technical problems because of their small size, the fact that DNA bases can slowly degrade over time, and the possibility of contamination with more recent DNA. Often such fragments have lost the chromatin structure associated with them in the living organism. Depending on the format, retrieval is more or less complicated and will be available at different stages of technological progress. The embalming process of organisms in amber, for instance, is still poorly understood at the chemical level. One point worth noting is that if DNA is available only in very minute amounts in extreme situations, this is not a problem if the bases are not too degraded. Powerful techniques exist, such as polymerase chain reaction, that can selectively and rapidly amplify small amounts of DNA to any desired level, basically by exploiting DNA's natural replication mechanisms. In time, it should also be possible to work with samples from which some of the DNA is missing, progressively increasing the amount of missing DNA a technology can tolerate. In the limit, one reaches the scenario explored in *Jurassic Park:* bringing back to life the dinosaurs that roamed the planet 65 million to 200 million years or so ago. In the film, dinosaur DNA is reconstructed from blood cells in the stomach of dinosaur-contemporary mosquitoes trapped in amber. We are still very far from being able to complete all the steps required for that example, but several first steps in that direction have already been taken.

For example, scientists are trying to bring mammoths back to life, a much easier task than bringing back *Tyrannosaurus rex,* because mammoths are more recent, having disappeared about 3,700 years ago on Wrangell Island in southeastern Alaska. One possibility would be to find a male mammoth that has remained frozen at low temperatures since his death. Sperm cells have particular properties that allow them to remain viable at low temperatures. So the idea would be to try to obtain mammoth sperm and inseminate a female elephant, most likely an Asian elephant because these are thought to be the closest living relatives of mammoths. This of course is a tall order, because it requires three things: (1) finding viable mammoth sperm; (2) being able to extract it and use it to artificially inseminate an elephant; and (3) ensuring that elephant and mammoth DNA and developmental programs are compatible enough to create a healthy hybrid—an elemoth, or rather a mammophant, because the father is a mammoth. With some luck, the mammophant will be female, so that the operation can be repeated a second time, and so forth. After three such operations, for instance, one would obtain an animal that is approximately 88% mammoth. All this is extremely difficult, but it can be planned and attempted even today. Alternatively, we could attempt to clone a mammoth directly from any well preserved cell by transferring its nucleus and associated genetic material into an elephant egg. This approach is also difficult and would require first polishing our skills at cloning elephants, which has not been done yet. In the meantime, the Japanese firm Field Co. Ltd. has sponsored a scientific expedition to Russia's Sakha Republic, led by reproductive biologist Akira Iritani of Kinki University in Japan, to hunt for a sperm sample from the remains of a woolly mammoth. The company hopes one day to set up some kind of "Mammoth Park." A French expedition led by Bernard Buigues found two frozen mammoths in 1998 on the remote Taimyr Peninsula and returned to the site for a similar attempt. In the fall of 1999, a twenty-two-ton block of tundra believed to contain the remains of a woolly mammoth was airlifted to a cavern hewed from ice, where scientists plan to thaw it.

Another case being discussed is the possibility of bringing back to life the Australian Tasmanian tiger. The last known living specimen died in a Tasmanian zoo in 1936, after a century of relentless hunting by early Australian settlers, who saw it as a threat to their livestock. Like much of Australia's other fauna, the Tasmanian tiger was an odd beast, more the size of a dog than a tiger, with a characteristic set of vertical stripes over its hindquarters. At least six Tasmanian tigers, preserved in alcohol in museums around the world, could perhaps provide a sustainable level of genetic diversity. As in the case of the mammoth, the hope is that the animal's DNA could be extracted and used to fertilize one of its closest living relatives, perhaps the Tasmanian devil. In September 1999, the state government of New South Wales announced the formation of a special government-private trust aimed at spurring research in this direction. Successful transplantation of a frozen embryo from one species to another closely related species has already been done—for instance, an ordinary house cat has given birth to an endangered African wildcat.

Thus, in time and with the proper technology, we will be able to clone any human being whose DNA is available in sufficient amount and viable form. Whereas the horizon of genetic immortality has been extended beyond Einstein, a suspicious degree of residual asymmetry is still left. What about human beings whose DNA has been entirely lost—for instance, through death in a fire? What about your mother-in-law, for whom you forgot to preserve a precious DNA sample? Could it be possible that she would be so unlucky as to have her genetic identity lost forever? As the reader may suspect by now, nature is not as unfair as it seems.

The sequence associated with any human genome, or for that matter the genome of any other organism, can in principle be generated in a computer. In fact, all possible DNA sequences of any length can be recursively generated in a computer. Of course, not all sequences correspond to viable organisms, and such a process is exceedingly slow. But this is irrelevant to the point I wish to make. The key point is that any given genome will appear in the list after some finite time.

In the case of humans, the search in genome space could be accelerated by using parallel computers, knowledge about the length of the genomes, and especially knowledge about genomes of all existing human beings. With advanced sequencing technology, in a few centuries it should be possible to rapidly sequence the genome of any living organism, and in particular of all living human beings. The statistical properties of the resulting sequences would greatly restrict the search space for the genomes of dead humans. For instance, in time and with powerful computers, one might be able to progressively reconstruct an approximate genealogical tree for humankind. After all, the number of all human beings past and present is only a few billion. I do not mean to hide the fact that this is a formidable computational task, without a well-defined unique solution, completely out of reach of today's technology. But at least some ideas of how to go about it exist, and it could become more tractable in a few centuries. If perfect, such a tree would yield all human genomes; if only approximate, it would at least considerably restrict the space of possible genomes that would have to be searched in order to hit all the genomes corresponding to humans having lived at some point during history. But this technical point of speed is really not central to our argument. What matters is that a human genome represents a finite amount of information and that it can be generated by a computer.

The last piece of necessary technology is to be able to clone a human being starting from the genomic DNA *sequence* alone. This is difficult because current disk drives are not a good packaging medium for DNA and cannot be inserted into biological cells. Although it cannot be ruled out that different hard drives, closer to wet drives and more compatible with biological processes and DNA packaging, may exist one day, a simpler solution is likely to become available in a not-too-distant future. First, one would find the human genome associated with an existing human being that is the closest to the genome to be cloned. The technology for this step is available but is still too costly and too slow to be applied on a human population scale. Second, one would culture a sufficient number of cells from this optimal human candidate. This technology is readily available today. Third, one would proceed with progressively modifying or editing all the bases (i.e., letters) in these cultured cells where the two genomes differ until a cell with the desired genome is obtained. After that, cloning can proceed as usual. The third step, selective editing on a genome scale, should become possible with future extensions of current technology. Considerable shortcuts should also be possible by pasting together portions of genomes from different existing human beings rather than editing a single genome. After all, this is in essence how genomes are built by nature.

Thus such technology is well within the fiction science extension of what is available today. Naturally, other approaches can be envisioned, such as learning more about how DNA is packaged in cells and how to reproduce the packaging process. We can synthesize oligonucleotides, relatively short DNA sequences. It is then "only" a matter of synthesizing a number of nucleotides covering the entire genome and progressively piecing them together and packaging them into chromatin, chromosomes, and so on. Thus there is little doubt in my mind that if technology continues to progress at the current pace, within at most a few hundred years it will be possible, if desired, to produce any human being with any given genome specified on a computer. This is true for past and present human beings, but also all virtual human beings; that is, all possible human beings, including those who have never existed throughout human history. This is in fact easier than what is required to bring back dinosaurs, because living human beings are currently plentiful and can provide living human genomes to start the process, living human egg cells, and living surrogate mothers if needed.

In the *Jurassic Park* scenario, the handicap is double and likely to require a gradual solution initially. To use a computer analogy, it is like having only fragments of a program and not quite the right hardware to run it. But what is perhaps most amazing about a genome is that it is a program which, when executed, builds the kind of computer that is necessary for the execution of itself! By progressively executing larger and larger portions of the program, one can build better and better computers for its execution. So for dinosaurs, one would have to start with DNA fragments and with whatever species is available at a given time that is the closest to the dinosaur to be cloned. In the case of dinosaurs and our current planet, this would most likely be a bird. This initial species can be used for an iterative bootstrap process. One would have to recursively iterate the process of reconstructing, synthesizing, and expressing larger and larger portions of the target genome. This is a form of DNA morphing.

## DNA Morphing

For millennia, humans have purposefully manipulated the DNA of many organisms, using breeding techniques and selection. These techniques are in some sense primitive and indirect. They are slow, crude, and in some cases difficult to reproduce, inasmuch as they rely on rare events that occur by chance. They correspond to randomly altering and copy-pasting a population of computer programs, executing all of them (for many organisms, execution can take several years), and retaining a subpopulation of programs with certain properties. Then the process must be started all over again with the selected subpopulation. In many ways, biotechnology can be viewed as a faster and more precise version of standard agricultural practice. Biotechnology is producing an array of techniques with unprecedented scope, speed, and precision that can operate directly at the molecular level. New variants of existing animals and plants are being engineered, from tomatoes capable of resisting rotting, to cows capable of producing human hormones. In the future, we will be able to edit genomes down to the level of single letters, and even now the effects of such editing are beginning to be understood scientifically. We will be able to create new species and to morph organisms in arbitrary directions (see figure 7.1).

It is remarkable to think that any living organism on the planet can in principle be morphed into any other organism by following a possibly long but finite sequence of relatively small DNA transformations. Not only is it possible to morph a man into a gorilla, but also into a crocodile, a fly, or, for that matter, an oak tree. To see this, consider two living organisms, A and B. To morph A into B, it is sufficient to produce small DNA transformations so as to go backward along the evolutionary tree that led to A until a common ancestor of A and B is reached. From this point one can proceed forward in evolutionary time with small DNA transformation steps until

B is reached. The more divergent the organisms, the longer the sequence of transformations produced by this algorithm. In fact, this algorithm does not guarantee that the morphing path found is the shortest possible. It shows only one possible path.[1]

It should be clear that this technique also applies to extinct species that could be re-created in this way, and to individual organisms within a species. It applies also to new branches of the evolutionary tree that could be created ad hoc, that is, completely new organisms or future descendants of currently existing organisms. All that is required at each step is the technology to selectively produce small modifications in a given genome, and this is becoming available. Because the modifications are small at each step, if the available technology requires it, adult females of the corresponding species can be produced to provide cloning eggs and surrogate mothers. With even more advanced technology, one should be able to dispense with the need for proceeding in small incremental steps. As we have seen, any genome could be listed on a computer. Perhaps one day the technology will be available to package the corresponding DNA sequence directly in wetware and, for multicellular organisms, initiate cell division and embryonic development directly, without having to find an egg of the same species and a surrogate mother. Ultimately, one may be able to directly synthesize the corresponding organism, although such advanced technologies are not a requirement for the arguments of this book. A slow, inefficient technology that can get the job done in principle is all we care about now; and it is not very far from what we already have. Full-blown DNA morphing is not yet available, but many of its components are in place and their integration is forthcoming.

Molecular bioengineering further challenges sexuality and sexual reproduction. Even with IVF, in vitro embryonic development, and cloning technology, we still must have some form of sexuality to produce substantial genetic diversity: at the very minimum, male sperm and female eggs. With molecular bioengineering, the need for a sexually based mechanism for generating genetic diversity is entirely removed. Diversity in genomes can be generated in the most direct way: simply by operating on the genomes themselves. This could be done both on a large scale—for instance, by chromosome shuffling—and on a finer scale by direct editing. By such techniques, human populations consisting entirely of individuals of the same sex could "reproduce" and even maintain or increase their diversity.

Some of the DNA morphing steps discussed, which involve going from an offspring to one of its parents or vice versa, can be considered coarse. Typically such steps conserve only 50% of the genes. It is clear that by manipulating DNA at the molecular level, much smaller morphing steps are possible, all the way down to the level of a single-letter mutation. Thus DNA morphing can be very fine-grained—essentially continuous for all practical matters. Continuous morphing brings the final challenge to our sexual identities by shearing it in the other direction. In one direction, we are being stretched away from the opposite sex because we would not need it to reproduce and diversify. We are also being stretched toward the opposite sex in a way that progressively blurs any differences. Any individual of a species can be continuously morphed into any other individual of the same species, including one of the opposite sex. In other words, it must be possible to interpolate almost continuously from any man to any woman at the DNA level. This is bound to challenge our still by and large binary view of sexes and their boundaries, and more generally all our taxonomic boundaries in biology.

It is essential to note that we have restricted our discussion of sexual identity to the genomic level. Our sexual identities also have something to do with our environment, the way we are raised, and the culture we live in. By bringing in these extra dimensions, the continuity and possibilities for interpolation are increased, not diminished. In other words, by raising the members of a population of clones in many different ways, one can produce a spectrum of different sexual identities that can be combined, so to speak, with the spectrum generated by progressively modifying the genomes themselves.

Molecular bioengineering breaks all taxonomic boundaries by creating new organisms and species in a way that would essentially be continuous at the DNA level. In this respect, such boundaries of living organisms are by nature flexible and blurry. Indeed, they are constantly undergoing revisions by biologists. Species, for instance, are defined as populations of closely related organisms that interbreed under *natural* conditions. With the advent of modern biotechnology, the notion of natural conditions will progressively lose some of its meaning. Interbreeding can and will occur more and more in the laboratory, in the test tube, and in the computer.

## Warping and Reversing Time

Our biological nature strongly influences our notion of time in at least three main complementary ways. First, at the physiological level we are inhabited by a number of rhythms and clocks, from the beats of our hearts to our circadian wake-sleep rhythms. A number of proteins have been discovered that act as molecular clocks and are responsible for some of the rhythms. Second, our time is also marked by biological development from conception to death. This is also in part controlled by our genes and by complex gene regulatory circuits. This human scale is expanded by our direct experience, typically, of five generations of humans, from our grandparents to our grandchildren, during our lives. Human history of course provides a further extension to this scale. Third, on a grander scale, we have knowledge of biological evolution through fossil and geological records, and through DNA sequence comparison. This evolutionary history is also ingrained in our genomes; hence the saying that ontogeny follows phylogeny.

All three notions of time can be warped through progress in molecular biotechnology. We have seen how cloning, molecular medicine, and other kinds of DNA alterations could significantly expand the life span. We can easily imagine new drugs that could interact with our subjective perception of time, with the speed at which memories are formed in the brain, and with all kinds of different molecular clocks. We could evolve organisms in artificially created worlds where the day-night rhythm is different from the one we live in. This is done in laboratories and in agriculture—for instance, to influence reproduction cycles in organisms as diverse as sea urchins, tobacco plants, electric fish, and trees. By programming environmental factors such as light, temperature, and water cycles, plants and animals can be tricked into behaving according to the corresponding artificial calendar period. We will definitely find drugs that interact with the developmental regulatory circuits or manipulate the corresponding genes in the direction of accelerating or slowing down development and life cycles. This is happening in the case of certain plants and livestock fed with growth hormones, where rapid growth is less costly from an agricultural perspective.

It is important to realize that developmental programs are exceedingly complex and that at least many decades of research will be required to understand them. In particular, these programs place significant but hidden constraints on gene manipulation and evolution. No matter what the ultimate function of a gene product is in an adult organism, if the gene interferes negatively with the developmental program, it cannot become part of the genome under natural conditions. Another example is the amplification factor: a small change during development can lead to large changes in final phenotype.

Molecular clocks have coevolved and are likely to have complex relationships. Imagine for the sake of the argument that we could find a way to accelerate human body development uniformly by a factor of 5 so that physical adult size is reached at the end of the fourth year of life. This would require us to learn how to walk by the third month and master the basics of speech by six months or so. But how can we be sure that the brain could adapt and learn the corresponding skills in such a short time? Perhaps several years of interaction with the environment and gradual synaptic changes are an absolute necessity for our current brains to master a language.

Finally, not only can time be distorted elastically, but in some strange sense it can also be reversed. This is another by-product of the generalized cloning technology. The day we can clone or realize any genome, we can put together organisms that during history and evolution were separated in time. With plain just-around-the-corner technology, we can have the clone of a son raising the clone of his father or mother—or grandfather, for that matter. By cloning the same person twice at different times, a family could be raising the equivalent of identical twins with different ages. Or one could have twins implanted in two different mothers. Incidentally, by impregnating the same mother with the same genetic embryo at different times, or by impregnating two mothers with identical twin embryos, we could study some of the most subtle effects of environment on development. With more advanced technology we can have the clone of Newton raising the clone of Einstein. With even more advances we could reintroduce Neanderthals or dinosaurs. In other words, biotechnology has the potential for genetically instantiating any form of life at time *t*, regardless of when it may have occurred during evolution or human history. In the next chapters we will encounter two other possible distortions of our notion of time, related to the cycle of life and death, and to the acceleration of information processing on Earth.

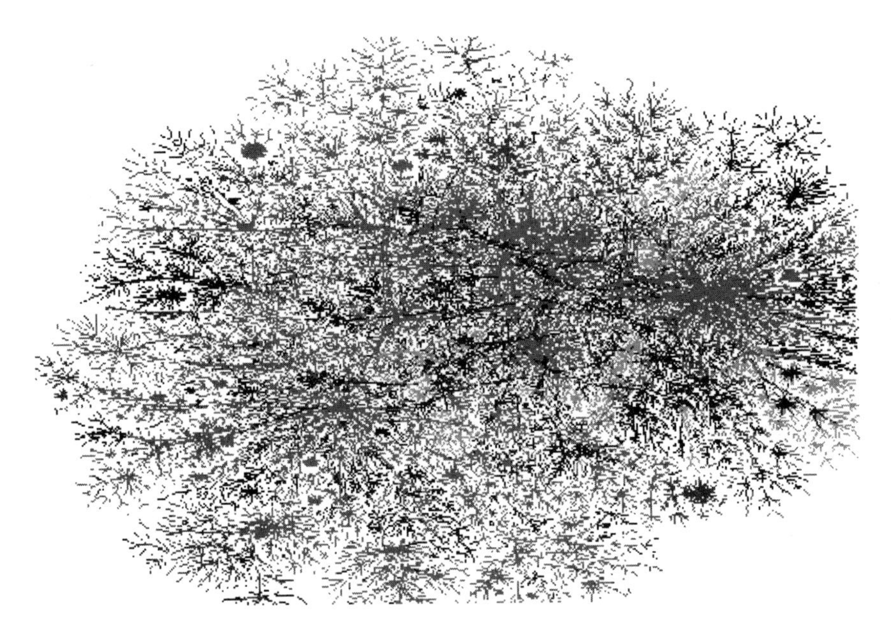

Figure 8.1 Cybergeography: large scale connectivity of the Internet. New connections are added to the Internet backbone every hour. Tracking and visualizing the topology of the Internet ecosystem is a challenging task. *Skitter* is a tool that has been developed to track global Internet topology. (Courtesy of K. Claffy and the Cooperative Association for Internet Data Analysis, San Diego Super Computer Center. Data collection and analysis: Daniel McRobb and Bradley Huffaker. Graph layout code: Bill Cheswick, Lucent/Bell Laboratories; and Hal Burch, Carnegie Mellon University. Work sponsored in part by the National Science Foundation and ARPA.)

# Computers

# 8

*We now have all the ingredients to build a distributed computing fabric which approaches science fiction.*
—David Farber, computer scientist

Before we proceed to the last fiction science scenarios and the human brain, we should briefly look at computers. Computers are essential in almost all areas of science, technology, and developed societies. But as they become more ubiquitous and the things they do for us become more complex, they begin to resemble, replace, or surpass us in many ways. More and more we are going to try to define ourselves with respect to machines. The frontier between biological computing and silicon computing has started to blur.

## Computations: Moore's Law

The essence of computers is to process information. One of the most striking facts about our computer technology is Moore's law, named for Gordon Moore, one of the founders of Intel, the largest manufacturer of computer chips. Moore's law is the driving force behind the modern computer expansion and provides a rough measure of how faint the distinction is between what computers and biological systems can compute. It simply states that the number of transistors or the number of operations that can be carried on a given piece of silicon real estate roughly doubles every eighteen months or so. Stated another way, computer power increases by a factor 10 every five years. The remarkable aspect of this law is that it was first noticed in the mid-1960s and remained true throughout the 1970s, 1980s, and 1990s. Today, we are reaching gigahertz computing clock speed and moving beyond it. There are of course fundamental physical limitations to how many transistors can be packed on a given surface area. After all, a transistor must be bigger than a single atom. But such limits have not yet been reached, although we are fast approaching them and the semiconductor industry is about to face its greatest challenge. Experts believe Moore's law will remain true for at least the first decade of the twenty-first century. A somewhat equivalent law applies also to computer storage devices in general, and to the volume of available data in many different areas, from biology to astronomy to finance. More generally, it is remarkable that the entire scientific enterprise has been following a kind of Moore's law for the past 300 years. Whether we measure it, for instance, by the number of scientists or the number of publications, science has roughly been doubling every fifteen to twenty years for the past three centuries. Most scientists who have ever lived are alive today.

Other computing technologies are being investigated and could in time yield different speed improvements. In addition to parallel computing, these include optical, quantum, and even DNA computing in which biological sequences are used to store information in the same way that they store genetic information. Computations are carried at the molecular level by cutting, copying, pasting, and so forth. More generally, any physical system that has a complex space of possible states, and in which we can exert some control over the dynamic trajectories, can conceivably be harnessed to perform computations. In fact, at the physical level, a computation is precisely that: a trajectory in the state space of a physical system. Thus, in principle, a computer could be built using billiard ball trajectories, although obviously this would not be practical. In the rest of this book we will use examples from silicon technology only. This is not to exclude other technologies, but rather because silicon computing is so pervasive and successful. Furthermore, details of a particular technology do not play a central role for the points to be made.

Computers already exceed human mental abilities in a number of ways, such as the precise storage and recall of large amounts of digital data and numerical calculations. Recursive algorithms to solve simple problems, such as the well-known Towers of Hanoi,[1] are easy to program and execute on a computer. The brain, on the other hand, has great difficulties with nested recursions a few levels deep. Even in some reasoning tasks, such as the game of chess, computers have now reached or surpassed the abilities of the most skilled players. In 1997, IBM's Deep Blue computer defeated Russian world champion chess player G. Kasparov for the first time. Many tasks remain, however, in which computers are making steady progress but still lag considerably behind humans. Typical examples are driving a car in a city, understanding speech, and carrying on a meaningful conversation on a variety of topics. One possible partial explanation for this lag is that the sheer computing power of the computer is still below the sheer computing power of the brain.

The best mainframe computers at the end of the twentieth century had a capacity of roughly 1 billion operations per second. What would be an estimate of the corresponding capacity for the brain? Though we do not know exactly the nature of the operations performed by the brain, we can estimate by counting the number of synaptic events per second. We have seen that the total number of neurons in the human brain is on the order of $10^{12}$. Considering that a typical neuron is connected to roughly 10,000 other neurons, the total number of synapses can be estimated at $10^{16}$. If we assume that on average a neuron produces ten signals—action potentials—per second, the total number of synaptic events per second in a human brain is on the order of $10^{17}$. Thus we need an improvement factor of about $10^8$ for computers to match the computational speed of the brain. According to Moore's law, this corresponds to about thirty-nine years. This is a very rough order-of-magnitude estimate, perhaps a little conservative with respect to other estimates that are closer to twenty-five years. It could easily vary by several years. It does not take into account variations in Moore's law, the possibility of using parallel computers, the fact that not all the brain is involved in solving a particular task, and a number of other minor factors. But the fact remains that according to most experts, the performance gap should disappear, and even begin to reverse, sometime between the years 2020 and 2050. The number of synaptic events could be off by several orders of magnitude, but as long as the exponential increase remains valid, a factor of 1000 adds only about fifteen years. Naturally, sheer computing power is a necessary component of intelligent systems but is not sufficient per se. After all, it takes many years of learning before humans can drive a car or converse on a variety of topics. How to use this power—the algorithms—is just as important, and there is still plenty to be learned at this level.

Single computers with large computational capacity are just one aspect of the emerging computing world. Perhaps an even more important phenomenon is the emergence of a large-scale distributed computing fabric. As chips become cheaper, smaller, and more capable, they can be inserted into all kinds of devices. The trend has already started with, for instance, appliances and cars, and is being extended in many directions, from telephones and TV sets to houses, clothes, and so on. Furthermore, in time all computing devices could be interconnected, and the global network to do it is being built now. The current Internet is just the first version, the progressive interconnection of all existing computing devices. Physical or wireless connections exist, for instance, between any two telephones or any two television sets with cables. Communications and computations, the transfer and processing of information, are becoming ubiquitous, distributed, and interconnected.

## Communications: The Internet

*You will need to work intensely to make the connection to information a new inalienable right, just like the right of assembly and the right for free speech.*[2]

—Eric Benhamou, CEO of 3Com, in a commencement address

*[The Internet is] mutable, nonlinear, interactive, multitasking, community and communications oriented, infinitely deep and mysterious—a bit like the female of our species.*

—Jim Banister, executive VP of Warner Bros. Online, a Time Warner AOL division, at a conference in June 1999

If the essence of computers is to process information, then the essence of the Internet is to transfer it. The fundamental idea behind the Internet is to provide a means of communication among all information-processing entities in the world. In its current implementation, it is limited in size, speed, geography, and the kinds of devices connected to it. These limitations have historical, technological, and economical reasons. But things are evolving rapidly, and the Internet is converging toward the asymptotic form it aspires to: (1) the transmission of information (2) as fast as possible (3) between any two processing points in the universe.

Today the Internet is limited to certain areas of our planet. There is little question that in the near future, one will be able to access it from any point on Earth and its atmosphere, if only by a wireless connection. Teledesic, for instance, is company that is building a global, broadband "Internet in the sky" using a constellation of low-Earth-orbit satellites. Backed by, among others, Bill Gates and Craig McCaw, Teledesic plans to deploy a network of about 300 satellites around Earth to provide telecommunications services, such as computer networking, broadband Internet access, high-quality voice and data, everywhere on the planet, including nonurban areas and less-developed countries. Iridium, a company with a similar strategy, was forced to shut down in 2000 for lack of sufficient revenues, after investing over $5 billion and deploying sixty-eight satellites. There is no reason why in the future the Internet could not be extended to other planets and other portions of the universe. The main limitation on a cosmic scale is, of course, the speed at which information can be transmitted, which cannot exceed the speed of light. On Earth, the speed of communication is still well below its theoretical limit but will continue to improve.

As briefly described in the historical notes in appendix D, the original cold war idea behind the Internet was to develop a communication system capable of withstanding a nuclear attack. As a result, one of the characteristics of the Internet is its distributed nature and complete absence of a central control. Packets of information traveling between computers do not follow a fixed path through the network. Rather, they are dynamically routed, depending on the origin and destination of the packets, but also on the state of the network, its level of congestion, and so forth.

The hardware embodiment of today's Internet is heterogeneous and partly overlaps with preexisting cable and telephone networks. Its backbone consists almost exclusively of optic fibers. These fibers can carry billion of bits per second, and the fibers of tomorrow will carry trillions of bits per second. New fibers are continuously being deployed within and across continents. The total data-transmission capacity across the United States is currently believed to more than quadruple every year, a rate significantly faster than Moore's law for computing power. Toward its periphery, the Internet hardware is more heterogeneous, ranging from older, twisted-pair copper telephone lines, to TV cable to the most modern wireless networks. In addition, it requires an extensive routing-switching infrastructure on which the fortunes of companies such as Cisco are built. The centrally planned telecommunications network originally built around circuit-switched voice traffic is now radically changing to a packet-switched distributed network for digital data capable of delivering video to the end users.

This heterogeneity has both historical and technological roots, but is also related to the fact that we are used to treating different types of information in different ways. By necessity, biological systems treat information acquired through different sensors in different ways, at least in the more peripheral areas of the brain. Yet in a sense digital information is a single quantity, regardless of how it is acquired or transmitted. Video, audio, and data are just that—information; and we ought to be able to manage, organize, personalize, search, store, edit, and communicate them just as we do for computer files. There is absolutely no reason, for instance, why in principle the Internet could not carry all telephone conversations or all radio and TV programs. In the same vein, there is no reason, other than historical, for not having our home PC control our TV set, our stereo, our refrigerator, our oven, our telephone, and any other appliance as necessary, and vice versa. In time, one should be able to access any radio or TV program, any CD, or any film in the world from any point on Earth almost instantaneously. This is already well under way for audio information because of its lower capacity requirements. One of the best-known new companies in this sector is RealNetworks, whose products allow one to download and listen to CD and radio music over the Internet, as well as build personalized digital jukeboxes. Video will be next.

The invention of the printing press and of photography created new media that were specialized for one kind of information and that were passive, information flowing directly from the page or the still image to the human eye. In contrast, the Internet is a new medium that affects all forms of information. Futhermore, it is "active" in the sense that all kinds of software can be embedded in the medium to actively control, search, modify, and filter information flows without human intervention. Messages can carry programs that can change the behavior of the receiving computer.

For the foreseeable future, the Internet will continue to grow rapidly with a homogeneous backbone and a more heterogeneous periphery that will also progress toward broadband communication rates; the current threshold to qualify for broadband is about 1 million bits per second. The periphery, the number of end points, will continue to grow exponentially and will be dominated progressively by wireless links. The main obstacle for broadband communication to individual homes is the so-called last mile challenge (see appendix D), increasing the two-way communication rates over the older telephone and cable infrastructures that were not designed for the Internet. In 1999, less than 1% of homes in the United States had broadband access to the Internet, but this is rapidly changing. In ten years, most homes are expected to have broadband connection and will be capable, for instance, of receiving the equivalent of ten movies simultaneously. A company such as Broadcom, which designs most of the chips supporting broadband communication across different technologies, stands a chance of playing for communication the role Intel has played for processing since the mid-1970s.

Thus the central tendency is toward a higher degree of transmission uniformity and toward having all forms of information (video, audio, and data) circulate through a single broadband information superhighway. In this sense, the Internet is rapidly becoming another kind of utility, similar to water and electric power. The information differences are relegated to the periphery—how to capture and render, how to code or decode, how to customize and personalize—and are not in the transmission. Therefore there is also no reason why the Internet could not also carry tactile or olfactory information, once the proper encoders and decoders are developed. Virtual reality gloves that provide tactile sensations exist now. To transmit an odor, for instance, one could imagine transmitting the chemical formulas of the various compounds it is made of, together with their concentrations. Once we have the hardware that can digitize a given type of information and the hardware that can re-create it for our senses, all it takes is transmitting a series of zeros and ones. There is no reason to do it differently for different modalities. The same is true for storage. We do not use compact discs for classical music, magnetic tapes for country music, and records for pop. Indeed, at the periphery of the Internet, at the interface between the Internet and humans, we are seeing the emergence of many devices—information appliances—with functions capable of speaking, browsing, E-mailing, and so forth.

In the early version of the Internet, most of the information-processing entities connected were almost exclusively silicon computers. But this of course does not have to be so, and is changing rapidly. TVs, portable telephones, and various handheld devices are connected, and some degree of convergence between laptops and portable phones is to be expected.

Cellular phones with video capabilities are becoming available. The Internet is being extended to processors in cars, in handheld devices, in home appliances, or in the clothes we wear, as well as to nonsilicon-based information-processing entities. Over 50 million information appliances are in use at the start of the third millennium. In fact, we can expect an explosive growth in machine-to-machine communication. A vending machine, a refrigerator, or a printer should be able automatically to notify a computer when it is running out of items, milk, or paper, by a wireless or other communication link.

The explosion in connectivity opens the door for completely new combinations of information streams, content, and interaction to which many of us are not accustomed and that create new horizons and challenges for mankind. Over the next ten years the amount of information that can reach our homes will grow by a factor of 1000. This will enable a new kind of interactive television in which, for instance, the distinction between conventional television and Internet browsing will be entirely blurred. Not only should one be able to request and receive the video information one wants, but one should be able to view it at any pace, to hyperlink it to other information related to its content in real time, to view parallel information, and so forth. Needless to say, the control of the flow of information reaching each living room will lead to new forms of family interactions.

Beyond the hardware and the sheer connectivity, however, the essence of the Internet, the largest of all libraries and databases, is its content and the forms of communication it enables. In the future, anyone on the other side of the planet will literally be as close to us as our next-door neighbor. In principle, we ought to be able to talk to, see, and even touch anyone in the world. Through small cameras located all over the planet, we should be able to view almost any location. A weak version of this is available today from sites that provide point-and-click access to photographs of a city, block by block. The Internet enables each one of us, as well as our organizations, to reach out in unprecedented ways; entertainment, education, and commerce, for example, are being deeply affected. New kind of searches are becoming possible. In principle, one ought to be able to hum a tune and receive the complete digital song through the Internet. And each one of us could be survived by his or her web site with a video of his or her life.

But transmission technology is currently ahead of regulators and creates ownership, security, and privacy problems. These problems should remain manageable, however. Privacy one day ought to become a knob that information-processing machines, including ourselves, wear and adjust with each communication. Similarly, the technology is ahead of the possible content and business models. A technology called MP3, for instance, combined with the Internet, allows computer communication, storage, and execution of any musical content in a way that completely bypasses conventional CDs and CD distribution channels. Soon the same will be technically possible for video.

Finally, the Internet affects computations. Communication between processing entities will allow for new forms of distributed computation on an unprecedented scale. Just think about harnessing the computing power of the tens of millions of computers that are currently idle across the world at night. Indeed, several projects of this kind exist, such as the SETI@home project. The corresponding web site states: "SETI@home is a scientific experiment that harnesses the power of hundreds of thousands of Internet-connected computers in the Search for Extraterrestrial Intelligence (SETI). You can participate by running a free program that downloads and analyzes radio telescope data. There's a small but captivating possibility that your computer will detect the faint murmur of a civilization beyond Earth." As of March 2, 2000, over 204,923 years of central processing unit time had been contributed to the project by almost 1.8 million users. Additional examples are given in appendix D.

The Internet and the broadband revolutions have arrived and are in the process of profoundly transforming all areas of our lives: from medicine to entertainment, from finance to politics, from manufacturing to commerce and education. Within twenty years, most electronic and other computing devices, including humans, will be connected anywhere, anytime. Potentially, any human will have the possibility to communicate and affect any other human. This most exciting, almost euphoric, and most prosperous era in history is occurring at an astounding rate: Moore's law for the Internet is "doubling every six months." But it is not only changing how we do things; it is changing who we are and how we view ourselves in the universe. Information technologies and biotechnologies are in the process of uncovering a new world.

## Artificial Life

*Within fifty to a hundred years, a new class of organisms is likely to emerge. These organisms will be artificial in the sense that they will originally be designed by humans. However, they will reproduce, and will evolve into something other than their initial form; they will be alive under any reasonable definition of the word.*[3]

—J. Doyne Farmer and A. Belin, computer scientists

Artificial life is one area of computer science that aims at creating and simulating life and living organisms on computers or other physical media, especially in terms of population dynamics and their evolution under simulated selection. Genetic algorithms and computer viruses can be viewed as two specialized domains related to artificial life. Genetic algorithms are optimization algorithms that simulate the evolutionary process. They can be used to search for optimal, or most fit, structures or programs in complex spaces, using the principles of Darwinian selection and survival of the fittest. They typically proceed by evolving a large initial population of more or less random suboptimal structures or programs over many generations through simulating operations such as sexual crossover, mutation, and selection. Computer viruses, like natural viruses, are parasitic programs that can not do much by themselves and need a host to become active. They can attach themselves to files and "infect" a computer over a network or from a drive. Once they are active, their execution can seriously harm the host itself—for instance, by deleting important information from memory or producing junk data that rapidly fill the available space or propagate over the network.

A typical artificial life simulation provides a software description of an environment, of the organisms contained in the environment, and of the interactions between the organisms and their environment. These interactions occur through a number of simulated behaviors, such as feeding, reproducing, and dying. Selective pressure can also be simulated by the introduction of fitness functions. A certain degree of randomness is often present in such simulations to take into account variability phenomena observed in the physical world. The execution of the program starts from a set of initial conditions and simulates the evolution of the system over a certain period of time. In general no rules specify the global population's behavior. Such behavior emerges from the simple individual rules and local interactions. Even on current computers, artificial life simulations can be slow, depending on the level of detail and the complexity of the interactions.

Two goals are implicit in the field of artificial life. One is to understand evolution and complex systems, and to find general principles of organization for theoretical biology. Artificial life systems expand our understanding of life-as-it-is, with its carbon-chain implementation on Earth, by placing it in the broader context of life-as-it-could-be. Simulated evolution, for instance, can be very different from natural evolution. For one thing, it must be faster. It also can be Lamarckian rather than Darwinian, in the sense that what is learned by an artificial organism can be embedded in its artificial genome if we wish.

The other goal of artificial life, more implicit, is to "play God," that is, to re-create life at the computational level and test how far one can get. But how far can artificial creatures go? And could things get out of hand—for instance, in the sense of human operators losing control of their creations? It is probably fair to say that in the short term there is little danger of that happening. First, the relative slowness of current computers leads to rudimentary results compared with natural organisms. Second, the organisms created so far by artificial life methods have had almost no direct interaction with or impact upon the physical world.[4] They exist only in the memory space of the computer used in the simulations, and become inactive when the power of that computer is turned off. But could that change?

In some sense the creatures of artificial life are not material; they are made of information and are similar to algorithms and mathematical objects. The computer language in which they are written, the operating system of the computer on which they are simulated, and, for that matter, the hardware of the computer on which they are simulated are in many ways completely irrelevant to their essence. These matters play a role only in the instantiation or embodiment of these creatures in a particular simulation. But how relevant is this to the essence of life? Could such a world be as alive as the one we currently inhabit? Could humans decide to inhabit such a world? Or could the world we inhabit be itself an artificial life simulation? We will touch on these issues in the next chapters.

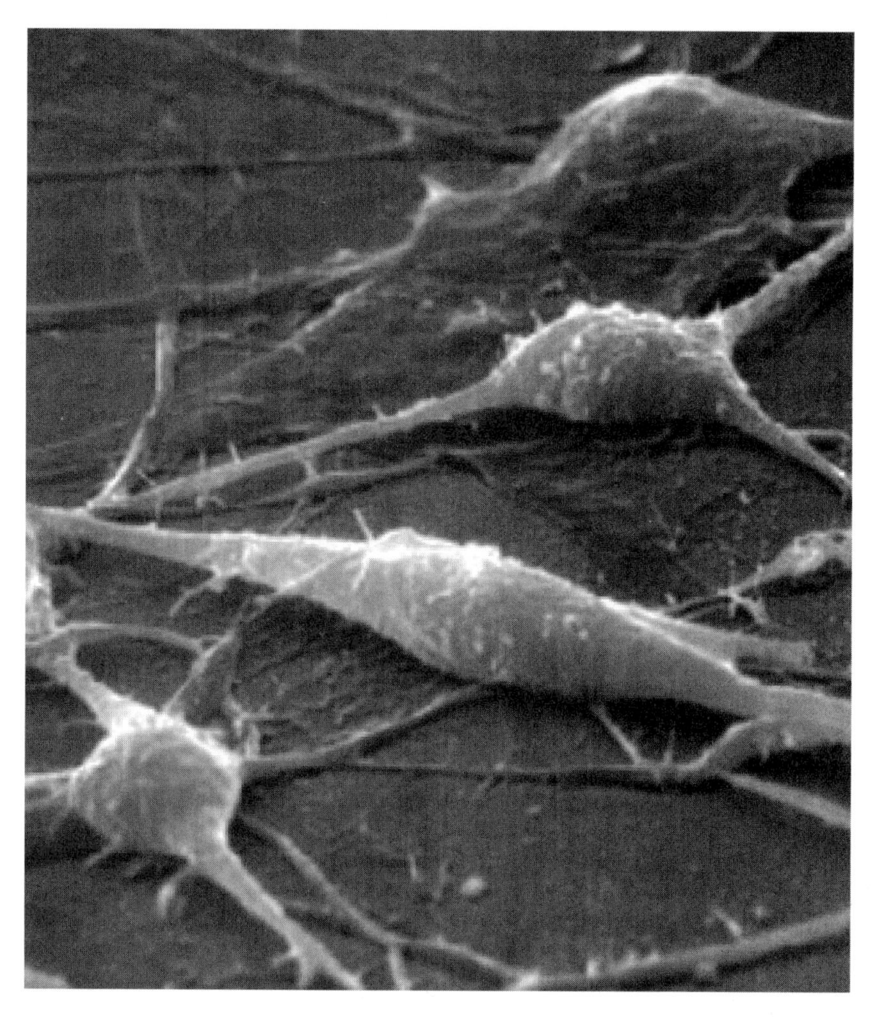

Figure 9.1 Neurons. (Courtesy of Dr. Scott Whittemore, University of Louisville.)

# The Last Frontier: The Brain

# 9

*The brain seems to be getting more complicated every day.*

—Francis Crick, physicist and biologist

The human brain of course plays a central role in who we are and in our notion of self. So far, however, we have not directly discussed any issues pertaining to the brain. The reason is simple: a substantial difference exists between the levels of maturity reached by molecular biology, on the one hand, and by neuroscience, on the other hand. To a large extent, we understand very little of the brain, although it is being investigated by many scientists from the molecular to the behavioral level, using an array of complementary approaches. Incremental progress is being made at all levels on the biochemical and biophysical properties of neurons and synapses, on the neuroanatomy and neurophysiology of neuronal circuits, and on their computational and behavioral properties. Powerful imaging techniques allow us to visualize the functional activity of the many areas of the brain under various conditions. But the overall impression is that we are far from understanding how we see and how we learn how to walk, let alone how we remember, reason, feel, and are aware.

In fact, this is an area of science where we have not found all the fundamental principles. Even basic questions about how information is precisely encoded into neuronal spike trains (for instance, how relevant is the precise timing of each spike?) are the subject of hot debates, let alone understanding the computations being performed. Neurons are exceedingly complex, nonlinear, adaptive dynamic cells. At a minimum, to understand a given neuron completely, we must describe all the other neurons with which it interacts, and this is a formidable task. Neuroanatomy is at a primitive stage in spite of a century of steady progress. In most cases, we do not know exactly what and how neurons compute, and there are so many of them, roughly as many as grains of sand on a beach.[1] In contrast, for almost any open question in molecular biology, the path to the answer is somewhat clear at the onset, although it may require long and tedious experiments. For some of the most difficult questions in neuroscience, such as the nature of consciousness, we are at a loss even to define the question precisely. This is in part because we do not have the right concepts.[2]

## Why Is Pain Painful?

*The chief wonder of all we have not touched upon yet. Wonder of wonders, though familiar even to boredom. So much with us that we forget it all our time. The eye sends, as we saw, into the cell-and-fibre forest of the brain throughout the waking day continual rhythmic streams of tiny, individual evanescent, electrical potentials. This throbbing streaming crowd of electrified shifting points in the spongework of the brain bears no obvious semblance in space-pattern, and even in temporal relations resembles but a little remotely the tiny two-dimensional upside-down picture of the outside world which the eyeball paints on the beginnings of its nerve-fibres to electrical storm. And the electrical storm so set up is one which affects a whole population of brain-cells. Electrical charges having in themselves not the faintest elements of the visual—having, for instance, nothing of "distance," "right-side-upness," no "vertical," nor "horizontal," nor "color," nor "brightness," nor "shadow," nor "roundness," nor "squareness," nor "contour," nor "transparency," nor "opacity," nor "near," nor "far," nor visual anything—yet conjure up all these. A shower of little electrical leaks conjures up for me, when I look, the landscape; the castle on the height, or when I look at him, my friend's face and how distant he is from me they tell me. Taking their word for it, I go forward and my other senses confirm that he is there.*

—Charles Scott Sherrington, neuroscientist, in *Man on His Nature* (1951)

For all its fascination, thinking about the brain always generates a sense of uneasiness. Aside from the sheer complexity of the subject, the uneasiness arises from the discrepancy that exists between our limited understanding of the brain at the biological level and most of our intuitive psychological concepts. Concepts we use to describe our mental life—words such as "intelligence," "memory," "self," "love," "anger," "free will," and "awareness"—were introduced into our language thousands of years ago when our ancestors where largely devoid of scientific knowledge. In fact, some of them were not associated with the brain, but were thought to inhabit particular organs, such as the heart or liver. When these words were introduced, our ancestors did not know anything about the real atomic[3] structure of matter, about DNA, about proteins, about neurons, and about computers.

These folk psychology concepts remain essential to our intuitive understanding of ourselves and our social interactions. But it is clear that they cover extremely complex phenomena at the neuronal level that are poorly understood. For instance, it is increasingly apparent that what we call memory is likely to be a constellation of different modules and processes associated with different anatomical areas and circuits, different populations of neurons, and different neurophysiological and biochemical dynamic processes, such as short-term and long-term potentiation of synapses, covering time scales from submillisecond to years. On a molecular scale, memory seems to involve not only biochemical changes but also morphological changes, such as the growth of spines on neurons. It may also involve changes in the DNA of the corresponding neurons to affect the expression of particular genes. At the functional level, evidence exists for many kinds of memory such as short-term and long-term memory, procedural memory (unconscious motor skills, such as riding a bicycle), and episodic memory (facts, one's biography). After a stroke or some other brain damage, some people selectively cannot recall the names of people, or are left with a memory span of only a few minutes or so.[4] As science progresses, the simple words of our ancestors will be replaced by a much larger number of more specific terms associated with different aspects of brain structure and function.

Nowhere is the uneasiness more apparent than with the problem of what philosophers call qualia, or the quality of our perceptions. In essence, the problem is the same for all our sensory modalities, but it is perhaps most clear with the perception of pain, due to its peculiar intensity and apparent "one-dimensionality." Imagine that a source of intense heat is placed near a point on your hand. What happens next is conceptually simple. A special family of dedicated receptors in your skin detects the heat. Their excitation is translated into tiny electrical impulses that travel through nerve fibers to the spinal cord. From there, the information splits into two streams of electrical activity. Local spinal circuitry rapidly produces an opposite stream of electrical impulses that travels back toward the muscles in your arm, causing them to contract and move the hand away from the heat source. In the meantime, another stream of tiny electric signals is sent up the spine to the brain, where they are processed: they propagate through neuronal circuitry, exciting or inhibiting other populations of neurons, and ultimately result in your sensation of pain. Most of these steps are so simple that they can easily be replicated with hardware.

We need only a robot arm, a heat sensor, a motor, and a processor. The processor can be programmed to reproduce the desired input-output behavior. When the temperature exceeds a certain threshold, the sensor sends a signal to the processor. The processor recognizes the signal and sends the appropriate electrical command back to the motor, causing the mechanical arm to retract from the heat source. We can even build a memory, store a few bits of information representing the fact that the corresponding stimulus should be classified as nocuous and avoided in the future every time it is recognized. But somewhere along this pathway a strange event happens in humans: we feel an intense sensation of pain, and it is doubtful that such sensation has been replicated in the plain robotic hardware. How can what on the surface seems mere transmission and processing of information—neuronal pulses traveling along different nerves and different areas of our nervous systems—be responsible for our subjective perception of something "painful"? And how does the pain we feel compare with the pain of someone else? These remain definite mysteries.

It is easy to assign a functional value to pain from an evolutionary perspective. It can be seen as a useful means of reinforcing important associations that are vital to preserve the life and integrity of an organism. In short, it is an anticipation of death that helps forge strong memories. For animals with planning and complex behavioral abilities, such as humans, a simple reflex of pain is inefficient. Knowledge of pain leads to better planning and avoidance, and hence to much better survival rates. In this sense, pain can be viewed as a first step toward self-awareness, and one could ask where the sense of pain arises first in the animal kingdom.

This provides a plausible explanation for why evolution may have found pain to be a useful mechanism. But it falls short of explaining its nature or mechanism, or explaining the "blueness" of the color blue. Similarly, in a broader context, it is fairly clear that our emotions play an important role in the way we process information. But it is not clear whether information could be processed as well without emotions—for instance, in silicon.[5] And if it cannot, how could emotions be emulated in or emerge from silicon computations? I do not have a good solution to these problems, but luckily we do not really need one for most of the points of this book. We can look at how computers and molecular biology question our notion of intelligence without having to provide a specific definition of something so complex and multifaceted. The questioning is so broadly based and coarse that it applies to our folk psychology notion of intelligence and any of its possible reasonable refinements. A more precise definition would be required to make precise predictions beyond our goal. But the uneasiness about the brain cannot be scooped under the rug entirely. It lurks behind all the issues that have to do with the notion of self and will resurface in later sections where fiction science begins to approach science fiction.

## The Limits and Expansion of Intelligence

*There are three events of equal importance. . . . Event one is the creation of the universe. It's a fairly important event. Event two is the appearance of Life. Life is a kind of organizing principle which one might argue against if one didn't understand enough—it shouldn't or couldn't happen on thermodynamic grounds. . . . And third, there's the appearance of artificial intelligence.*

—Edward Fredkin, computer scientist

Developments in computer technology and biotechnology raise a number of puzzling questions about intelligence and other human faculties, even if precise scientific concepts covering them are not yet available. A complex faculty such as intelligence is under the influence of both genetic and environmental factors—nature and nurture. Most of the debate focuses on the relative weight of these factors, and biology is on the verge of reaching the point at which both can be assessed scientifically. But from the point of view adopted here, the exact relative weight is not really central. What is central is that we could manipulate both factors, and therefore we should be able to manipulate intelligence, or any other faculty, regardless of precise boundaries.

At the level of nature, we have seen that we could clone the Albert Einsteins as well as the Marie Curies. We could produce entire populations of them and/or mix their genomes in arbitrary fashions. We could also use cloning technology to learn how to tease out natural from environmental influences in a more scientific way. With or without cloning, we will progressively learn about the complex genes involved in the development of the brain and that influence intelligence. We will certainly develop molecules, drugs, and other techniques that can influence the brain, promote repair, and improve functions such as memory storage and recall. Good memory is probably a prerequisite for good intelligence.

As our understanding of biological intelligence deepens and begins to explode into many different facets, we could learn how to bring more change to a particular facet. We may find along the road that other things, which traditionally have been dissociated from it, such as emotions and/or social skills, influence what we call intelligence. These in turn are to some degree dependent on genetic factors, and with time could be acted on at the DNA level or by using particular drugs. In general, we could learn how to select and modify multiple genes to produce genomes with a potential for greater intelligence. In time, we could progressively create new beings—not necessarily *Homo sapiens*—with higher forms of intelligence. Perhaps from *Homo sapiens* we could derive beings with bigger brains, or even with multiple brains. Finally, it is conceivable that one day we could learn how to grow functional nerve tissue in vitro and hook up different wetware modules, and perhaps build new computers out of biological components (figure 9.2).

On the nurture side, we can learn how to manipulate environments in complex ways. We can first manipulate environments, using genetics again, by grouping particular combinations of clones together. We could clone Einstein (figure 9.3), and his parents, siblings, and friends if we wanted to re-create some of his environment. We could raise 1000 Newtons with 1000 Einsteins and/or let them work together. Science and technology could allow us to study and optimize our natural environment and lifestyles, from exercise, to exposure to pollutants, to nutrition.

Figure 9.2 Two-headed Pacific gopher snake (*Pituophis melanoleucus catenifer*). (Courtesy of Visuals Unlimited)

With computers, multimedia, and virtual realities we could also progressively command arbitrary environments. It is conceivable that human brains could be raised in manipulated or artificial environments to develop certain properties. Some of this technology is available and in fact has already been used in elementary ways. Kittens have been raised in artificial worlds containing, for instance, only vertical or horizontal stripes, to study the effect of the environment on the development of the visual system. Our education system could also be viewed from this perspective: in fact, the notion of a natural environment for the brain is not well defined at all. The Internet is a gigantic environment for our brains that is bound to exert an influence over time on how we learn, how we access information, and how we communicate.

Whereas computers have begun to create new virtual environments of various sorts, they have an even more direct impact on our notion of intelligence through their sheer number-crunching power. Electronic computers have been around since the late 1940s. Yet their progress has been staggering, and shows no sign of slowing down. As a result of Moore's law, the PCs that sit on our desks are several orders of magnitude more

Figure 9.3 Albert Einstein, person of the century, on the cover of the December 31, 1999 issue of *Time*. (Courtesy of *Time*.)

powerful than the first vacuum tube computers that occupied entire buildings after World War II. In the 1990s Deep Blue, a computer built by IBM, was the first to win a chess competition against Garry Kasparov, the human world champion. IBM recently announced plans to build Blue Gene, a supercomputer consisting of more than 1 million processors, each capable of 1 billion operations per second (also called 1 petaflop). Blue Gene would be over 1000 times more powerful than Deep Blue and would initially be used to predict how proteins fold into the three-dimensional shapes that allow them to orchestrate life at the molecular level.[6]

In the chess match, Deep Blue did not use the same algorithms used by the brain of Kasparov, which remain largely unknown. As expected, Deep Blue's approach was more combinatorial, rapidly sorting through large numbers of carefully selected possibilities. But does it really matter? After all, planes fly without flapping their wings. How much intelligence comes from algorithms and how much comes from hardware is to some extent a moot point for this discussion. Often both go together, and progress is continuously being made on both fronts. Algorithms that run too slowly on today's computers may suddenly become practical in twenty years. Progress in both hardware and software is beginning to make automated speech recognition possible. By the mid-twenty-first century we should witness the emergence of computers capable of performing a number of operations per second that exceeds the ability of a single human brain. What lies in such a world of higher intelligence? What kinds of algorithms or "intelligence" can be implemented in such computers? What kinds of mathematical theories could they develop? And how will we harness such power?[7]

In the first half of the twentieth century, before the advent of modern computers, the mathematician Alan Turing laid the foundations for several fundamental concepts in computer science, such as what it means to be computable by a machine. More informally, he also devised a famous intelligence test that now bears his name. According to the Turing test, a machine is intelligent if it is capable of carrying on a conversation with a human in such a way that the human cannot detect whether he or she is talking to a machine or a human being. The point of the test is not in the use of speech. In fact, in its original version the conversation is to be carried by typed messages that are blindly exchanged between the two parties, an E-mail conversation if you will. Machines that do a reasonable job of speech recognition and speech synthesis exist, and there is little question that the residual problems in this area will be solved in the relatively near future. The real difficulty of the Turing test lies, rather, in the semantic aspects of human conversation—being able to understand the *meaning* behind sentences—in order to sustain a sensible conversation over a broad array of topics.

For well over half a century the Turing test has retained a sort of mystique that still clutters our anthropomorphic thinking. Surely it is useful and captures our imagination. But scientifically it is also ill defined: the parameters of the "conversation" should be more precise. In time, there will be different levels of "Turing proficiency," including the ability to crack new jokes, write a novel, or generate new scientific theories, all of which are implicit in the ability to carry on varied conversations. How humans pass the tests is highly unclear. Again, what is the balance between hardware and software? Are sensory inputs from the external world essential to learn the meaning of words and converse about the world? Alternatively, is it sufficient to implement a sort of large algebraic system capturing the internal relationships between words? After all, in a dictionary each word is defined in terms of other words. And if one were to need external inputs, would one also need emotions and sensations to learn the real meaning of things? Or would a camera and microphone suffice? How can a finite piece of software generate or understand the infinite number of possible sentences of a language?[8]

The human brain is a living proof that a physical system can be built which passes the Turing test. It is only a question of time before machines will be able to pass the test in one of its forms, and embryonic versions of them are mushrooming everywhere. Programs exist that can "converse" within a very limited domain. Programs also can generate new mathematical proofs within a limited axiomatic setting. And tons of expert systems can suggest new hypotheses or diagnoses in a variety of complex situations, sometimes with a degree of expertise, in a restricted domain, that are equivalent to or exceed those of the human expert.

This foggy situation is exactly like that of chess in the 1960s and 1970s, when people were wondering if computers could "think," or be "creative" and beat a human chess player. After all, what are the algorithms used by human chess players? Perhaps machines could be created that beat the average human player. But what about a world master, like the legendary Bobby

Fischer? At the beginning of the twenty-first century we clearly see that machines can thrash humans at chess. They can come up with new strategies that surprise us or would seem creative to a human expert. In addition, this can be achieved by using brute force algorithms that almost surely are quite different from what humans do, but yield similar or even better results. With hindsight, it was only a matter of having the right amount of computational power and the memory to match it. It cannot be done with a machine capable of only 1000 elementary operations per second. But when the number goes up to a billion or more, it becomes a relatively easy task, another large number effect. It is not easy to guess exactly how much computational power is necessary to pass the Turing test and when that level will be reached, but the usual back-of-the-envelope calculations suggest it will be well within a few decades. Human intelligence is a relatively recent phenomenon, only a few thousand generations old, and it has not been doubling its capacity at each generation, as computers do. One day it might be viewed as a historically interesting, albeit peripheral, special case of machine intelligence.

Finally, there is the interface between brains and silicon. Up to now, we have used very slow mechanical interfaces such as punch cards and keyboards, or even voice. Mechanical communication could expand further if computers could learn how to interpret our body language, track the movements of our eyes, take our pulse, measure the conductivity of our skin, and so forth. To some extent, this is implemented in the computer gear that assists some people who are paralyzed by accident or disease, such as physicist Stephen Hawking (another good candidate for cloning), who has Lou Gehrig's disease (amyotrophic lateral sclerosis). But mechanical communication is bound to have a low bandwidth. Can we imagine a more direct interface between silicon and neurons with a higher bandwidth? Could neurons and transistors talk to each other directly?[9]

Nothing in our science and technology would lead one to think that such a form of communication is not possible. In fact, it already exists in a primitive form. Neurons and brain areas can be stimulated electrically and neuronal electrical activity recorded. This is routinely done in neurophysiological laboratories and by neurosurgeons. Sensations, memories, and even movements can be elicited in this crude way. In university laboratories, neurons have been grown on chips, and the first artificial cochleas, retinas, and even noses have been developed.[10] The first brain and spine silicon chip implants are under experimentation. We can then begin to think about the possibilities of growing nerve tissue and brains in vitro and connecting them to silicon devices. Of course, initial experiments can be carried out with simple and reproducible nervous systems such as those of invertebrate organisms that have a fixed set of a few hundred neurons with a rigid pattern of programmed connections.

Like genetic engineering, bioengineering or the mixing of hardware with wetware is a disturbing idea. And like genetic engineering, it is not at all new in human history. Prostheses and tools of all kinds, from eyeglasses and hearing aids to screwdrivers, are precisely that: hardware components created to overcome the limitations or deficits of our bodies.[11] Even computers could be viewed as brain prostheses. And consider the case of a very precise operation performed under microscope by a surgeon with the help of a robot harm receiving commands over the Internet. The man-machine boundary has already become very thin.

What is most disturbing is the interface, the degree of proximity and integration between human and machine, wetware and hardware. As long as the interface is distant—fingers against a keyboard, eyeglasses sitting on a nose—the concept is not too disturbing. Yet many examples of more intimate interfaces exist. We have artificial knees, hips, shoulders, and heart pumps implanted in living humans. In vitro fertilization is another example of a very intimate interface. The distinction between humans and robots is likely to continue to blur. Thus artificial cochleas, retinas, and noses, which are all in the works, and corresponding implants should not be terribly surprising. Human experiments with cochlear and retinal implants are already in progress. In some form, these should become commonly available well within a few generations. But what about implanted processors that enhance our senses or our motor abilities? And what about the possibility of implanting processors higher up in the processing hierarchies of the brain?

It is difficult to imagine what can be done with intelligence and other faculties several orders of magnitude beyond our own. Almost by definition, what can be done is in a blind spot that our brains cannot really see. All we can say is that one comes away with the strong impression that a vast uncharted territory of "higher" intelligence exists which we can only dream about and are on the verge of exploring. What works of music or mathematics can be produced by large computational systems, orders of magnitude beyond the single human brain? Because of evolution, education, and our physical reality, we have come to think of our brains in terms of entities that are by and large isolated, with precise boundaries and distinct from each other. This does not have to be so. In time these isolated entities will become more and more "open" and networkable, and will process information in tandem with other information-processing systems, whether brains with similar structure or different structure, or silicon-based or other processors.[12]

## Sizing the Self Up

Without getting caught in philosophical issues, most of us would agree that we are the products of at least two components: our genomes and our personal and social history. We have seen that the raw information corresponding to a genome sequence is rather simple and is storable on a single hard drive. But what about our history? Is it definable? Can we store it? Can we reproduce it or bring it back?

At this point it is useful to make another order-of-magnitude calculation and ask how much information with respect to the external world goes in and out of a human brain throughout a lifetime. Note that there is also a great deal of communication going on at each second within the brain itself, or between the brain and the internal organs. This communication and processing impacts the external world, of course, but only indirectly, by modulating the activity of our motor system.[13]

On the input side, the brain receives information about the external world from our sensors in the eyes, ears, nose, mouth, and skin. Consider a single eye. The eye is connected to the brain by the optic nerve. The optic nerve contains on the order of 1 million nerve fibers or axons, projecting from the retina to a relay station called the lateral geniculate nucleus, and then to the first cortical visual areas in the back of the head. Visual information is encoded into spike trains traversing these nerve fibers. The number of spikes per second on each fiber is highly variable, but on average has an order of magnitude of 10 to 100. Neuronal spike duration is typically a few milliseconds. If we consider spikes as all-or-none events, on the order of 100 bits per second are traveling on each fiber.[14]

With 1 million nerve fibers and 100 bits per second on each axon, the retina is sending at most 100 million bits of information per second to the brain. Whereas this is a formidable number, substantially larger than the bandwidth of telephone or conventional TV transmission, it corresponds to only 12 megabytes per second, given that a byte is equal to 8 bits. Such a rate is well below the transfer rate of our fastest optical fibers. Now we have two eyes, and although our visual system seems to play a predominant role, we have other senses. To account for that, let us multiply the number above by a factor of 10 and say that a rough estimate for the number of bits the brain receives about the external world from our sensors is on the order of 1 billion per second. So one minute of input would easily fit on a current 8-gigabyte (GB) PC hard drive.

The lifetime of a seventy-year-old individual contains about 2.2 billion seconds (not distinguishing between wake and sleep; presumably less information is transmitted from the world to the brain during sleep). Therefore a generous estimate of the total number of bits received by the brain in a lifetime is 2.2 followed by eighteen zeros: $2.2 \times 10^{18}$ bits, approximately equal to $2.7 \times 10^{17}$ bytes. This is less than the total data traffic carried by AT&T over a period of a few months. It also corresponds to the total memory capacity of 100 million PCs, each with a 2.7-GB hard drive. For comparison, it is estimated that there are over 400 million PCs on Earth at the beginning of the twenty-first century. In other words, the lifetime inputs and outputs of the brains of several individuals can easily be stored on the entire Internet today. If this seems a staggering amount, consider the following. First, this is an overestimate, and not all these bits are likely to be relevant. More important, Moore's law applies not only to computing but also to memory. In fact, in recent years memory has grown slightly faster than computing power, doubling every year or so. We can currently store 10 GB per square inch of memory real estate, and 40 GB per square inch should be available within two years. At such a rate, it will take approximately twenty-seven years—only a generation—to have common PCs with sufficient memory to store a lifetime of human brain inputs.

Now what about the brain outputs? The only output of the brain to the external world[15] is through our motor system, that is, through our muscles. Output nerves are separated into those which go to the head and those which descend the spinal cord. It is intuitively reasonable to guess that the number of outputs is substantially lower than the number of inputs; controlling muscle fibers seems simpler and coarser than carrying detailed visual information. Indeed, the ratio of inputs to outputs is about 4 or 5 to 1. If for simplicity we estimate the total number of motor fibers at about 1 million, we get estimates that are one order of magnitude lower than those given above. So the motor output per second corresponds roughly to 100 million bits, with the same storage consequences. Incidentally, it is important to keep in mind the feedback loop between output and input; what we see depends on where we direct our eyes, which in turn depends on what we see, and so forth. In conclusion, all the inputs and outputs of a human brain over a lifetime correspond to a finite amount of information that would fit on the combined memory of a large number of computers and, most likely, will be storable inside the memory of a single computer in a few decades.

Let us now define the "external self" as consisting essentially of a complete genomic sequence together with the recordings of all the inputs and outputs of an individual's brain over a lifetime. The main point of the calculation above is to show that the external self is a legitimate quantity—a large but finite amount of information—storable in a computer memory and as immortal as any other string of bits. The genomic component of the external self is measurable today, although at an impractical cost. The technology to obtain the input-output component is not available today, or only in a very primitive form. We can simultaneously record several neurons, but the methods are both invasive and limited to a maximum of 100 or so channels. To a certain extent, we can also record rough global activity over entire sensory or motor areas, using imaging or other techniques. Although recording all inputs and outputs precisely remains a formidable task, it is feasible in principle and techniques will continue to improve. In fact, a very good approximation is essentially available today by recording external events themselves (images, sound, temperature, movements, etc.) rather than corresponding neuronal events.

Indeed, we could use cameras and microphones to record an approximation to all the things that an individual sees and hears throughout a lifetime. This information could be augmented by using temperature sensors, and even sensors that register input not available to our biological senses could provide additional information regarding the environment surrounding the life of this individual. It could be further complemented with, for instance, historical, genealogical, and biographical information. Similarly, at the motor level we could get an approximation by filming all the actions of a person and/or by putting many position sensors along the body.

If such information were available, you would probably have the feeling of knowing the person very well, although for obvious reasons you would not be able to review all the information in real time and would be forced to sample it very sparsely. In fact, for other people you do not have access to much more than this form of external self. It is only when thinking about yourself that you have the strong impression that there is more to you than your external self, and hence infer that the same must be true for others. First, the external self does not give us direct access to a person's emotions and internal thoughts. Second, it does not explicitly tell us what the behavior of the individual would have been under a set of circumstances different than the one experienced, what she would have done if . . .—free will, in other words. Finally, the personal sense of self—self-awareness—also seems absent from the external self.

We can try to size up the "internal self" along the same lines by imagining that we could somehow record the electrical activity along the internal connections in the brain. A rough calculation gives something on the order of $10^{27}$ bits over a lifetime (appendix E). Whereas $10^{27}$ is a formidable number, a billion of billions of billions, in memory size it corresponds only to fifty-six years of technological progress if memory storage of typical computers were to continue to double every year, starting from a current level of about $10^{10}$ bits. By this I do not mean at all to imply that we will be able to record this version of the internal self before the end of the twenty-first century. It actually seems safe to bet against such a feat. For one thing, it is not at all clear that computer memory sizes will continue to grow exponentially beyond a horizon of two or three decades. After all, there are fundamental physical limitations to packing with current chip technologies. More fundamentally, perhaps, being able simultaneously to record the signals emitted by each neuron in a noninvasive way is a daunting task that is very far away from what we can do today. But what matters is that such definition of the internal self is conceivable and yields a large but finite bound. The slow process of trying to develop the corresponding technology has started, for instance, by simultaneously recording electrical activity from multiple neurons, by studying the electrical activity of neurons that are grown directly on chips, by developing various brain imaging techniques, and by trying to record the internal self of simple organisms with a small number of well-identified neurons.

Yet even a lifetime recording of neuronal electrical activity is unlikely to yield a complete, satisfactory picture of the self. Evidence is mounting that as a result of environmental and genetic influences as well as of learning, the brain is constantly modifying its computational structure. This can include phenomena as diverse as growing new neurons, axons, and dendrites; changing their morphology (e.g., by growing protuberances called spines); changing the chemical composition of synapses (e.g., increasing the density of certain types of receptors); and modifying the DNA of neurons or regulating the expression of different genes in different cells. These phenomena introduce several additional layers of complexity, which incidentally could further delay the date at which computers will catch up with brains in terms of raw operations per second. If what matters, besides the electrical signals between neurons, are these fine-grained structural and chemical properties of neurons and synapses, the quantity we ought to look at can be approximated by a study of all the proteins contained in a brain, because these are what is involved in signaling and gene control. If we could take a snapshot of all the proteins in the brain, say every second over a lifetime, we would obtain a record bounded by an order of $10^{31}$ bits. If, to get down to the level of DNA and single ions, we wanted a snapshot of all the atoms in a brain, say every millisecond over a lifetime, we would have something in the range of $10^{40}$ bits (appendix E).

In summary, the external self is a relatively easy to define quantity that can be approximated with current technology. It can be estimated to lie within a comfortable range of $10^{18}$ bits, a size undoubtedly within reach of computer storage technology. The internal self is also a well bounded entity. Yet it brings us into a more uncomfortable and uncertain range of $10^{27}$ bits and beyond, where speculation remains difficult and the required technology is daunting, even if its embryonic form is already developing today.

## Immortality Revisited

Through the use of technologies such as DNA sequencing and cloning, our genomes become essentially immortal, in at least two different ways. First, as long as human or other operators are available to perform cloning operations, a genome can be executed and expanded into a living human being. This provides a kind of immortality conditional on the survival of humankind itself, its technology, and the willingness to perform cloning. Second, and perhaps more important, a genome can be viewed as a string of letters or a program with a finite amount of information. As such, it belongs to the same space where all mathematical objects and algorithms belong—the information space, as we called it in chapter 8. In this sense, a genome is as immortal as a square or a circle, regardless of whether or not it has a physical incarnation—that is, regardless of the hardware to store or run it. This is also true of the external and internal selves. Genomes and external selves, for any human or information-processing device, could be stored and made eternal over the Internet. Yet, the genome, the internal selves, and the external selves in their still digital format are likely to appear somewhat unsatisfactory to those who have an interest in immortality.[16]

Thus the next question concerns the degree to which our physical bodies and brains can approach immortality. Consider the body first, with the exception of the brain. For the sake of argument, imagine a population of selfish, narcissistic, and immoral humans interested solely in perpetuating their physical bodies. Here again cloning is particularly handy. By farming clones, each individual in the population has available an unlimited supply of fresh, young, compatible organs. Undoubtedly, from a purely technical standpoint, the organs of a properly raised healthy clone are about the best possible candidates for transplantation with little risk of rejection. By such immoral techniques, we now have an immortal liver or an immortal kidney. Of course we have many other more ethical options by which the same results could be attempted, such as building better artificial organs, bioengineering better animal organs, developing organs in culture using embryonic stem cells, and using stem cells and other techniques to directly promote regeneration.[17] Regeneration in biological systems is possible, even to amazing degrees. Salamanders are capable of regenerating entire severed limbs and spinal cords, as well as portions of their brains, at any age and as many times as needed. There is absolutely no reason to believe that with the proper manipulations of gene regulation and expression, such capacity cannot be awakened in human cells.

Many laboratories are working on bioengineered animal organs or artificial organs and tissues, such as artificial hearts, heart valves, pancreas, knees, bone, teeth, and skin. Several products and prototypes are on the market, and some have been available for quite a while. Thousands of people have received artificial knees, and millions have artificial teeth. The company Minimed, for instance, has developed and successfully commercialized artificial insulin pumps and glucose sensors for diabetic patients. The ultimate goal in this area is to combine a sensor with a pump to build the equivalent of an artificial pancreas. Human cornea equivalents have been artificially grown in the laboratory.[18] By growing cells on a biodegradable scaffolding, heart valves have been grown in the laboratory and are being tested in lambs. Unlike existing alternative animal or mechanical valves, such bioengineered valves could have the advantage of being able to grow with the recipient. Furthermore, they may not require the administration of blood-thinning drugs to avoid blood clots, as mechanical valves do. It is therefore not too difficult to accept the idea of having an immortal liver, heart, or kidney. But what about the brain?

Cloning, gene therapies, advanced molecular medicine,[19] and surgical procedures such as organ transplantation, together with a better understanding and control of environmental factors, can render our bodies essentially immortal. The same techniques have the potential for greatly extending the brain's longevity: its ability to fight off diseases, to resist aging, to prevent neuronal degeneration, and even to regenerate. For several decades, it was thought that nerves and neurons in the brain and spinal cord could not regenerate in the adult. But an exception has been known for a long time: olfactory neurons in the olfactory epithelium regenerate themselves on a regular basis, probably to counter noxious effects of continuous chemical stimulation from the outside world. It is becoming clear that with the presence of proper factors and/or with proper gene, cell, and tissue manipulations, regeneration in the central nervous system is possible.[20] In fact, evidence shows that a small area of the brain involved in memory produces new neurons into old age, as do other areas of the cortex in primates.[21] Furthermore, stem cells injected into the spinal cord of paralyzed rats have been used to restore partial function. Several research laboratories are working on these problems. Techniques are also being developed by which drugs could be delivered directly to specific areas of the brain. Yet, it is not clear exactly how far all these techniques could extend the life of a brain. It is fairly certain that the "natural limit" of 120 years can be reached, but what about beyond that?

Regeneration by itself has the potential for extending the brain's life indefinitely, although sufficiently precise control of regeneration is a formidable problem to address.[22] Brain regeneration also shows that our notion of self must be continuous rather than discrete: as neurons or axons regenerate, the brain changes; hence, so does our notion of self. If the changes are small, the self must change continuously, as it does from day to day under myriad external influences. One could imagine a very slow regeneration process, almost imperceptible, in which small anatomical changes are integrated into the self through the many adaptive changes neurons are capable of. If the changes are substantial, however, changes in the self could be more abrupt. In any case, selective regeneration could progressively lead to a form of morphing, in which one brain could be morphed toward a different one. Needless to say, we are extremely far from such technology, from being able selectively to control neural regeneration and "brain sculpting," or to understand the complex balance of brains that would have to be respected by such manipulations. But we do have the tools and knowledge to begin exploring such directions—for instance, by examining the genes and proteins involved in the development of nervous systems. Other techniques, such as head transplants as well as various forms of freezing, can also be considered in the context of prolonging the life of bodies and brains.[23]

If immortality may not be impossible, as some scientists are starting to believe, what consequences would it have on human societies, besides the extinction or at least redefinition of life insurance companies? Our first naive reaction is to project ourselves into such a world and wonder, for instance, whether we would be motivated to have children, whether life would become boring, and whether some would still enjoy risky activities such as bullfighting. But immortality technology is more advanced than human cloning or in vitro babies. By the time it is achieved, humans and their societies would have changed considerably, so that it does not make much sense to slightly extrapolate from our own. We can speculate, however, that in a society where immortality is common, the need for each individual to procreate would be greatly reduced and that such a society would have the means to create whatever it calls humans by techniques more sophisticated than the cumbersome approach utilized by evolution. In an environment with limited resources, an issue of control over the creation of new "humans" would have to arise. And death, albeit not inevitable, could still play a role in this society, in particular in connection with new births. Perhaps, at least during a certain phase, such a society would have to impose a limit on how long each citizen could live. Almost surely, the notions of life and death and their perception would be very different from our own.

Behind the issues of self and brain immortality lies another, very difficult problem, that of whether a brain or a self could be copied or reproduced with very high fidelity. And if so, is there a limit on how close the copy can be? Granted that these questions require a more precise definition, it is clear that any attempt at reproduction could be carried out by using either similar hardware (another brain) or completely different hardware (e.g., silicon chips). Thus I will first try to ask a more specific question and then address it with different hardware approaches.

Before addressing these issues, however, note that producing two similar brains at the same time $t$ does not help with the problem of increasing the life of the self. On the other hand, being able to produce at a future time $t_2$ a brain similar to a given brain at a previous time $t_1$ has the potential to satisfy the requirement for immortality that some of us may have. In fact, such technology would do much more. Let us imagine that a given brain at time $t$ can evolve in the future in many possible directions, depending on chance or noise, variability in the inputs, free will, and so on. One can follow a sort of cone of trajectories or paths starting from any time $t$. Up to now, each of us has followed only one path inside such a cone. But if the technology above became available, it is clear that one could start sampling the cone along both the axis of time and the axis of variability among the paths. And this could be done in real time or after a delay. In this way, the snapshot of our brain at time $t$ becomes immortal and reproducible at any future time, and potentially so do all the snapshots of our possible brain at times greater than $t$. This is true not only for snapshots along the path the brain actually followed, but also for snapshots as it would have evolved under arbitrary different conditions—that is, all the virtual persons and selves we could have become but did not, because of the trajectory followed.

## The Distance between Selves—The "Organic" Approach

To consider these issues, we must now address the problem of comparing brains or selves in order to decide how similar two such entities really are. Given the brain's complexity, it seems extremely unlikely that two perfectly identical ones could ever be produced. In fact, this is probably true of simple cells if we require identity at the level of single atoms or proteins. Even if possible, such identity would probably last only an instant, because random fluctuations and variable conditions would cause complex molecular assemblies to diverge. As pointed out, however, such an identity requirement is too stringent and unnecessary. It is trivially violated by two identical PCs of the same brand and model. Yet, for all practical purposes, they perform the same computations in the same way. What distinguishes them at best in an interconnected world is the fact that they have different IP addresses.[24] But then how can we measure similarity in a meaningful way that is computationally relevant and avoids atom-by-atom comparison, and what degree of similarity can we achieve?

A position one could adopt is that each individual is absolutely unique and irreproducible, and that the "distance" between any two human beings is essentially infinite. This, however, runs contrary even to our most basic intuitions, experiences, and folk psychology that humans are comparable and some are more similar than others. In fact, intuitively it is quite clear how one could begin to construct such a distance[25] and how it could be refined in the future. Even today, we can take as our starting point the notion of external self. We can conceive how to define a distance at the level of single genes and progressively extend it to entire genomes. We will encounter many technical hurdles along the way—how to weigh genes and their relative locations, how to handle noncoding regions, and so forth—but all seem quite surmountable. The distance between two genomes should become relatively easy to estimate once full genome sequencing technology becomes widespread. We could augment such distance with any anatomical and physiological differences we could measure.[26] More important, we could start defining a computational component of the distance based on input and output functions—basically everything heard, seen, and done by the two brains over any period of time.

To compare two information-processing devices, we compare their input and output streams of bits. Many techniques in computer science and information theory are available to compare streams of bits: how much information they contain, how related they are, how they can be compressed, and so on. For simplicity, they are not discussed here, but it is clear that one could begin assessing the distance between two input-output functions. If nothing else, one could approximate the distance by using the input and output streams over a truncated reasonable period of time—for instance, while undergoing a standard battery of tests constructed by a committee of psychologists, psychophysicists, and other experts. In summary, it is reasonable to suspect that in time, with the appropriate massive amount of work, a meaningful distance over external selves could be defined, and in fact we already have the means to approximate such a distance.

With a plausible distance we can begin the process by which similar brains can be constructed at different times and then assessed. Suppose we start with a given brain at time $t$. Ideally we should have as much information as possible about it: this would include the underlying genome, information about anatomy, and information about the history up to time $t$, possibly including a recording of all the input and output streams of information up to time $t$. To produce a similar brain at a later time, one could start with a clone to guarantee genomic and gross anatomical similarity. The clone could then be submitted to a stream of inputs that resembles the original one as much as possible. Depending on the advancement of the technology, this would involve constructing a sort of gigantic movie set to re-create an environment as similar as possible to the original one, including cloning all the relatives and friends of the individual to be duplicated and re-creating their own environments. Though this seems an almost endless process, it could in principle be carried out with increasing degrees of precision. If it is hard for you to picture it, start from easier cases such as children, or even pets, living at almost the same time (i.e., $t_2$ close to $t_1$). How similar can two two-year-olds be, if genetically identical parents, grandparents, and other relatives raise them in identical socioeconomic conditions, with genetically identical siblings, in the same house, neighborhood, and city at almost the same historical time? Furthermore, the same experiment could be carried out in parallel a large number of times to compensate for natural fluctuations and ensure at least one success. Though one guesses that the reproduction could never be perfect due to chance events, in time we will have the technology to control both underlying genetics and the environment with an increasing degree of precision. Regardless of whether or not this is desirable, it should become feasible to produce approximations of a given brain that, without being identical, are similar in terms of external selves on an unprecedented scale. At such a scale, the space of external selves appears potentially dense and continuous, as opposed to discrete and sparsely populated.[27]

This approach to duplicating a human brain is based on the use of another biological brain, with the same genetics, together with the natural learning abilities of brains in similar environments. Hence it is "organic"—with no offense intended, of course, to the defenders of organic products. Alternatively, we can also explore how duplication could be attempted using different, more artificial, substrates.

## Machine Learning and Reverse-Engineering the Brain—The "Artificial" Approach

Human brains are physical systems of such flexibility and complexity that any reasonable artificial approach we can conceive today to approximate them relies heavily on some form of learning. In other words, it is difficult to conceive of a technology capable of assembling, entirely from scratch, a brainlike information-processing system without undergoing any extensive learning phase. Nature obviously uses learning very extensively to produce mature human brains. In fact, learning together with cloning is what we have used so far to try to approximate a given external self the organic way. But even today learning can be embodied in software and in hardware that is nonbiological. And this raises the question of whether external selves could be approximated by using other substrates together with the extreme fringes of a computer science discipline called machine learning.

Machine learning is the study of learning from a computational point of view, what is learnable and what is not, together with designing hardware and software learning systems to solve different problems. Software learning systems are complex programs that can learn or modify themselves when presented with particular information in order to achieve certain tasks. In particular, they can be designed and trained to learn a given input-output behavior from a set of examples. They typically contain a large number of parameters that can be adjusted during training to optimize a particular fitness or error function. After each series of examples, algorithms exist to modify the parameters in order to decrease the error. Artificial neural networks are a class of such learning systems among many other possible ones. These networks typically consist of interconnected artificial neurons that compute by applying a nonlinear function to the weighted average of their inputs. They were introduced in the 1940s to approximate biological neurons with weights representing synaptic strength of corresponding connections. Today we know that biological neurons are significantly more complex than the current artificial neurons. In spite of this fact, it can be shown that essentially any reasonable function can be implemented by a network of artificial neurons. The fundamental issue of course is the size or complexity of the corresponding network and how to find it, how to determine its topological structure and the value of the weights on the connections. Artificial neural networks with as many as 100,000 parameters have been successfully trained to address problems ranging from speech or handwriting recognition, to detecting credit card fraud, to controlling chemical plants, to analyzing DNA and protein sequences in bioinformatics. Even for problems for which reasonable solutions have been painstakingly derived over the years without learning methods, machine learning approaches can often derive comparable solutions in very little time.

Learning from examples is a statistical inference or model-fitting problem. The fundamental problem is being able to generalize, to produce the correct behavior on examples not previously seen. A simple lookup table that memorizes all the training examples behaves perfectly on the training set. On the other hand, such a program has no generalization abilities and therefore is useless. Is there enough information in a training set to learn the corresponding task to a reasonable degree of precision and certainty? And if so, can we learn it efficiently, and how? How do we decompose a complex learning system into modules? These are some of the questions scientists in this field are studying.[28]

The question, then, is whether or not the complex input-output functions of our brains are learnable. A plausible argument can be made that we are living proof that it is possible. This is not to say that learning such functions is not an extremely complex problem which may take us many centuries, if not millennia, to solve. But the rough path toward its solution is clear, and the effort to reverse-engineer the brain has begun.

One of the old and wise sayings in biology, first used for relatively simple objects such as molecules, is that function follows structure. This remains largely true for more complex systems such as organs, and for the brain. If we are to understand how brains work, we need to continue to work on brain anatomy, brain biochemistry, and brain physiology at many different spatial and temporal scales for quite some time into the future. In fact, valuable information in this area can be obtained by combining molecular biology and neurobiology to study brain development when different types of receptors, neurons, layers, and areas are created and laid out: to study function, study structure; and to study structure, study development. Numerous additional insights are also being produced through more systemic and cognitive approaches, such as brain imaging techniques and computational approaches directed, for instance, at detailed modeling of single neurons or synapses. Although little more than the surface has been scratched so far, all these efforts have produced invaluable insights.

In less than three pounds of electrified jelly, the brain packs a trillion neurons, each contacting as many 100,000 other neurons through tiny fibers. Electrical signals as well as chemicals race down this jungle of fibers and across synaptic clefts, converging and diverging throughout a recurrent dynamic network of staggering complexity that is responsible for all our thoughts, memories, and emotions. The first lesson about the macroscopic structure of the brain is that it is organized in a vast network of recursively interconnected and hierarchically organized modules. A good example is the visual system of primates, for which several dozen modules have been discovered. Some of them, for instance, are involved in processing the location of objects or their motion ("where"), whereas others are involved in processing the nature of objects in a visual scene, regardless of their spatial coordinates and orientation ("what"). Similarly, memory seems to include a number of different modules, although we are far from having a good sense of its overall organization. This coarse decomposition is likely to be important if we want to reverse-engineer the brain and is likely to be reflected in the modular decomposition of learning systems designed for brain approximation. The modules themselves contain complex circuits of many different types of cells organized in different anatomical structures, such as layers. Although neuronal circuits are particularly complex and messy, they are also very sparse. The typical cortical pyramidal cell, the main type of neuron in our cortex, is on average connected to 10,000 to 100,000 other cells—a very large number indeed, but far from the $10^{12}$ one would have in a fully interconnected brain. This is true at all scales, including the macroscopic modules described above. Thus hierarchically organized, recursive sparse networks are a ubiquitous feature of brain organization.

We have seen that artificial neurons are orders of magnitude simpler than biological neurons, but also that they have universal approximation properties. So let us assume that a biological neuron is the computational equivalent of $10^5$ artificial neurons. This gives a brain-equivalent artificial neural network with $10^{17}$ artificial neurons. In addition, let us assume that the typical artificial neuron is connected on average to $10^5$ other artificial neurons. This number is consistent both with having a fully connected network of at most $10^5$ artificial neurons representing a biological neuron, and with a biological neuron being on average connected to another set of $10^5$ biological neurons. Thus we end up with something on the order of $10^{22}$ connection weights in this gigantic brain-equivalent artificial neural network. The main point is not the fact that this number is gigantic, or that it could be off by several orders of magnitude. The main point is that such a calculation can be considered and in all cases yields a finite, albeit very large, answer. The largest artificial networks people have trained to this date have $10^5$ to $10^6$ parameters. For the sake of the argument, imagine that the size of the largest artificial network we know how to train doubles every year. Then an easy calculation shows that starting from the current level, a century is enough to reach networks of size $10^{22}$! Undoubtedly, as discussed in chapter 8 and in appendix E, this is an overly optimistic estimate, and if this project is ever carried out, it will require a much longer period of time.

But one cannot underestimate human ingenuity applied to this problem over several centuries. In many ways such a program has already started out from many different angles, with scientists working on the anatomy, chemistry, and physiology of real brains; on artificial learning systems, neural and other; and on their interfaces and relationships. Progress is being made both on simple invertebrate nervous systems containing a fixed, small number of precisely defined neurons, and on complex ones such as the visual system of primates. If this duplication project could be completed for humans and result, say, in a chip, this chip would certainly be able to pass a Turing test and do much more. Small variations in the chip would also result in a nebula of information-processing devices with similar but nonidentical characteristics and external selves.

## The New Music

*Entre dos Aguas.*

—Paco de Lucia, flamenco guitarist

A computer scientist friend who used to play music professionally was recently complaining to me that it was becoming difficult to produce new music. Traditional Western music has only twelve notes, and many ways of combining them in pleasurable, and not so pleasurable, ways have already been explored. Specific music styles are even more constrained, from Baroque classical music to popular modern music. Blues, for instance, is built around only three chords, arranged in very specific rhythmic patterns. Similarly, flamenco harmony is built around four chords. We owe a tribute to the remarkable ingenuity of musicians if original blues and flamenco music is still being created.

The twelve notes in question can be played at different frequencies, in different combinations, and with many different instruments, although this is more a way of providing new colors than of generating new melodies. In some cultures, instruments are used with a different set of notes. Originally, even in Western music the same flat and sharp notes used to be slightly different.[29] It is also true that musicians have explored composition rules that go beyond classical harmony. For instance, in dodecaphonic music, the twelve notes are arranged with a more or less uniform distribution rather than the skewed distribution associated with a particular key. Musicians are also working with new sounds and noises. In this regard computers, with their increasing abilities to store and alter sounds, and add effects and distortions of all kinds, have become important elements of the musical landscape. But the fact remains that the melodic space can sometimes feel crowded. And part of this musical expansion results from musicians seeking to overcome the limitations of conventional music. These limitations are perhaps less obvious in the case of images and shapes. To some extent, this is probably a consequence of the fact that music is essentially one-dimensional—musical events are linked to one-dimensional time. Two-dimensional images and three-dimensional shapes are perceived as having more degrees of freedom. How, then, can we open the musical space?

Music results from the processing by our auditory system of the vibrations of air molecules. These vibrations are generated by particular boundary conditions determined by the vibrations of musical instruments, loudspeakers, and, more generally the acoustic environment. Therefore, if we want to open up the musical space, we basically have two possibilities. We can work on the boundary conditions, that is, on composers, performers, and instruments. Or we can try to work on the auditory system.[30]

At the production level, musical talent is a human faculty that is not without parallel with intelligence. It probably depends on a combination of genetic and environmental factors, and just like intelligence, we have the potential for manipulating both. We could try to manipulate genes and genomes to produce extremely talented composers and performers. In the case of performers in particular, it is clear that many genes, other than those controlling the auditory system, could be involved, such as genes controlling the size and shape of the hand, the motor system, and so on. Biofeedback principles, for instance, have been used to manipulate the environment and improve how violinists control vibrato with more efficient muscle movements. Vibrato is produced by a vibration of the fingers over the strings and is one of the most technically difficult things to master on the violin. Stretching these ideas, one could conceivably try to produce new types of instrumentalists—for instance, a pianist with four hands, or a guitarist with twelve fingers. Could artificial Siamese twins—with one brain for each hand—become particularly good performers? New instruments could also be created. The Internet is an ideal medium for transmitting any amounts of digital music over practically any distance. Musicians are already trying to play together in real time over the Internet.[31] Most likely, the Internet will one day foster the creation of gigantic concerts (TV does already that) and jam sessions of planetary if not cosmic proportions.

More important, perhaps, progress in computers enables us to tailor increasingly complex boundary conditions. Music can be represented, composed, and recorded in digital media in many ways, from standard sheet notation, to sound waves, to the physical modeling of particular instruments. The digital record can be edited and manipulated down to the level of single bits, just like genomic DNA.[32] Sound can then be synthesized back through a synthesizer, but also in many other possible ways, including some that are not yet available but could be realized within the next century, such as reproducing the movement and pressure of a bow on a violin. In fact, if I had to make a prediction, it would be that music will become increasingly computerized and its mechanical aspects will progressively disappear. Within a few hundred years, traditional instrumentalists could vanish almost entirely, and something derived from the computers of today could become a kind of universal instrument connected to the Internet.

But the direction where the musical space could really be opened up is the human auditory system. Needless to say, the greatest degree of freedom is achieved by manipulating both the auditory system and the sound space in a coordinated fashion. Some psychedelic drugs alter our perception of music by acting on our brains. At the nurture level, virtual realities, computer games, and special earphones could be designed to help shape auditory circuits. More important, over the long run our genes could be manipulated to expand our auditory space, that is, expand the range of audible frequencies and the resolution of frequencies, intensities, tones, and timing of auditory events. We know that such feats are possible in biological circuits because they exist in other species. Dogs, for instance, can resolve high frequencies that are inaudible to us. Bats have a sonar navigation system that can resolve exceedingly small time intervals, less than a millionth of a second. Barn owls can localize and capture prey with high precision in total darkness by processing minute sound differences associated with their asymmetrical frontal ears. Music could thus be written with more than twelve notes, a wider range of frequencies, and a wider range of rhythms and effects. Music also has a strong emotional component, and therefore one could consider acting on the corresponding brain centers, too. And what kind of music could we process and appreciate if we had two integrated brains and/or four ears?

Conceivably we could also find ways of providing direct stimulation to the auditory nerve or the auditory cortex, or even interface auditory neuronal circuits with silicon ones. The vibration of air molecules would then become unnecessary. On the output side, could we find a way of directly playing the music we have in our heads? In experiments on animals, the optic nerve has been redirected toward auditory areas and vice versa during development. Artificial silicon cochleas have been designed. Could computers listen to music? Ultimately we come back to the question of qualia, as we would have with any other sense, which is beyond our scope.

Interestingly, we still do not know why humans love music and seem to need it. Its survival advantage is highly unclear,[33] and none of the theories advanced so far is very convincing. But answering these difficult questions is not necessary to see clearly that we have plenty of room, if desirable, to use technology for expanding our current musical space in several directions and relieve some of us from the boredom of twelve-note melodies.

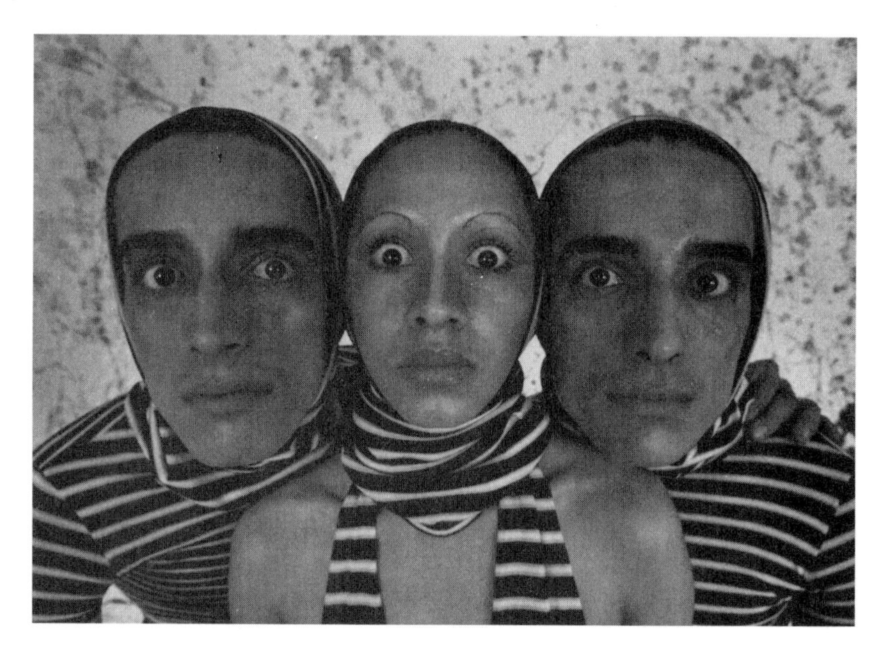

Figure 10.1 "Morales Twins and Priscilla." Photo by Gilles Larrain, originally published in *Idols* (New York and London: Links, 1973). (Courtesy of Gilles Larrain Studio.)

# Ethics and What Can Go Wrong

# 10

*In the end, America is not ruled by ethics. It is ruled by law.*

—Steen Willadsen, biologist, in the *New York Times,* December 2, 1997

*Consider—an animal that is extinct, and is brought back to life, is for all practical purposes not an animal at all. It can't have any rights. It's already extinct. So if it exists, it can only be something* we have made. *We made it, we patent it, we own it.*

—Michael Crichton, in *The Lost World*

We all have a sense that farming humans is an appalling idea. The point of this book is not to recommend that fiction science experiments be carried out, but rather to try to understand their implications for who we are. Yet there is tension between the natural curiosity and enthusiasm of scientists for new discoveries, on the one hand, and repulsion toward the instrumentalization of humans, on the other. Not to mention that if humans belong to a continuum, it becomes difficult in the long run to forbid for humans, or quasi humans, experiments that are not forbidden for other animals. Thus, although ethical matters are outside the main scope of this book, they cannot be underestimated and are briefly surveyed in this chapter.

## Ethical Foundations

*If that [animal cloning] works, and even if detailed adjustments have to be made for other species or types of cells, there seems to be no barrier in principle to the possibility that scientists could eventually make an identical copy of you from just a scraping of skin or a tiny snippet of your spleen. We hardly dare to think of the ethical difficulties such achievements would bring in their wake.*

—Rob DeSalle and David Lindley, in *The Science of Jurassic Park and the Lost World* (1998)

*The possibility of controlling our children's genetic destiny strikes me as only good. It is grossly unfair that some families' lives are dominated by the horrors of genetic disease.*

—James Watson, biologist, in a special issue of *Time,* "The New Age of Discovery" (Winter 1997–1998)

Many bioethics texts share the same conservative punch line: we ought to be extremely careful and proceed very slowly with biotechnology, because we must preserve our notion of humanity and of who we are. Of course, I agree that we ought to be careful. But preserving our humanity is not a good reason. It amounts to saying that we should be careful exploring the oceans, because we have to preserve our notion that the Earth is flat. But flat the Earth is not. We should be careful because the oceans are dangerous. But we may well have to abandon our concept of a flat Earth if we want to align ourselves with reality. The point of view adopted here is more radical and is *ontological* rather than ethical. It contends that our common notion of who we are is erroneous in the same way that the notion of a flat Earth is wrong, regardless of our personal feelings, desires, and ethical positions. The way we see ourselves is in part the result of historical and evolutionary accidents. It does *not* correspond to the broader reality that is rapidly emerging with the explosion of the Internet, computers, and biotechnologies.

---

### *Dogmatic Approaches*

Historically, ethical thinking in Western societies has been strongly influenced by religious traditions and the notions of human nature and human dignity. From this perspective, the supreme goal of ethics is to establish universal principles of right and wrong, and what constitutes moral behavior. Implicit in this approach is the belief in the existence of a stable human nature, and a view of eternal human values and humanity that must be preserved and fostered. Such a foundation may have seemed plausible during the last few thousand years of history, when human nature was relatively stable. But it should be clear now that this was a mirage and sometimes the source of mistakes, such as the justification of slavery on the grounds that slaves were not human. The idea that things stop where we are and should be preserved as such is entirely unjustified. Things do not stop where we are, or where we think we are, but extend continuously. It is difficult to use who we are as an absolute foundation for ethics if we do not *know* who we are, and even more so if who we are is rapidly shifting. In particular, the notion of human nature is *not* a stable one unless we artificially try to freeze it, which is highly unlikely to succeed. Such a concept is likely to evolve substantially in the future, if only because of progress in computer technology and biotechnology. As a result, ethics will have to become more dynamic and more closely mixed with law, politics, and the decision-making processes.

The second notion—human dignity—was influentially developed by, among others, the German philosopher Immanuel Kant in the 1700s. In a nutshell, human dignity for Kant signifies that man must always be an end, never a means—or at least not in an exclusive way. This tradition is evident in the first international document aimed at establishing general guidelines for human rights, medicine, and biological research—the Nuremberg Code of 1947, partly motivated by the atrocities of World War II. The Nuremberg Code was closely followed by the 1948 Universal Declaration of Human Rights of the United Nations. The first article of the declaration stipulates that "All human beings are born free and equal in *dignity* and rights. They are endowed with reason and conscience and should act towards one another in a spirit of brotherhood." Since then, the concept of human dignity has been central to a number of official documents and conferences attempting to establish a bioethical framework. But dignity is difficult to define and rely on. Is assisted suicide a form of dignity-preserving murder?

Whereas we can all strongly sympathize with the concepts of humanity and dignity, their extremely vague definitions are bound to be problematic and inefficient when the time comes to apply them with legislative precision to situations raised by modern biotechnology: from gene patents, to abortion, to IVF, to artificial twinning, to cloning, to name the mildest ones. In a number of cases the debate has boiled down to the thorny question of the status of the embryo—is it human or not? Does it have dignity or not? It is difficult, and in my judgment even impossible, to reach a consistent position across these issues if one's guidance comes solely from the vague notion of human dignity. It is hard to defend the position that the embryo is human. Aside from the practical fact that this would stiffen medical research and contradict common practices of abortion and IVF,[1] the embryo does not have a brain and therefore is devoid of emotions and consciousness. Yet, if it is not human, does it mean that everything should be allowed? In fact, the problem can be reduced to that of a single cell, the equivalent of the egg just after it has been fertilized. On the one hand, this single cell is just that—a complex assembly of molecules devoid of any notion of self. As such, one should be able to manipulate it in any possible way, as we routinely do many other types of cells. On the other hand, in less than nine months the same cell can evolve into a human being that deserves all the human rights we cherish.[2]

All dogmatic attempts to escape from this conundrum seem somewhat baroque and illogical. For instance, the attribute of humanity could be extended to everything that has the potential to become human. Dignity could be said to be respected when life is promoted rather than destroyed. But then abortion immediately becomes unethical and cloning ethical. One could invoke the concept of "natural," as in natural reproduction. But then antibiotics are unnatural; so is the doubling of our life expectancies over the last century. And aren't humans used at least in part in blood transfusions or organ transplantions? One ends up being forced to add one clause after another to deal with each practical case one by one, defeating the purpose of a dogmatic approach. People who are afraid of science and cloning and fond of human dignity should ask themselves how a society could logically subscribe to the death penalty—the ultimate use of a human as a means—and at the same time proclaim the absolute dignity of human life.

Traditionally, law represents the implementation of ethics within a set of given socioeconomic constraints. Laws often tend both to channel and to adapt to technological changes. Laws, in particular, implicitly or explicitly rely on definitions and boundaries between notions such as human and nonhuman, human and animal, man and woman, alive and dead, genome closeness (incest), and so on—all boundaries that are going to be somewhat fuzzy and in flux over time. One obvious area where law has been forced to adapt rapidly, and will continue to do so, is the area of patents.[3] The original patenting framework was strongly influenced by the absolute view of ethics and the world: only works of humans are patentable (inventions); works of nature (discoveries) are not. Thus mathematical theorems and laws of nature were not patentable—they belong to all mankind. But algorithms, software, and now genes and genomes have challenged this framework and are creating a number of patenting nightmares that are difficult to resolve and require, in my opinion, a substantial modification of the current patenting system. Should one be allowed to patent genes? Genomes? Are strings of bits the work of man or of nature? What happens when computers can pass the Patent Office version of the Turing test?

At one level, as we have seen, genes and genomes are like algorithms or mathematical formulas and belong to the information space. The first thing to observe is that our patent laws are *not* uniform across the information space. Mathematical theorems and proofs cannot be patented in principle, but algorithms and genes can. This lack of uniformity is the result of historical and economic accidents, and is bound to become more and more an area of tension as the boundary between natural and artifical continues to thin. Gene patents were authorized for economic reasons, to justify the substantial investments made by pharmaceutical companies, for instance, to hunt down genes responsible for particular diseases and to develop corresponding drugs to cure them. With the race between public and private sectors to complete the human genome, however, gene patents have come under fire. Finding the DNA sequence of a gene is much easier than finding its function. Should knowledge of the sequence alone be sufficient to file for a patent? And if not, how much of the function ought to be known?[4] Should a single company, such as Celera Genomics, be allowed to patent a large fraction of important human genes?

Genomes are in some sense collections of genes, and, on this basis, there is no absolute reason why they should not become patentable at some point. After all, sequencing all the existing bacteria and viruses, or developing entire new organisms with particularly useful functions, is also bound to require substantial investments. And on this basis one could justify such patents. In fact, genetically modified organisms have been ruled to be patentable by courts because they are not products of nature. It is hard for me to see how such an obsolete distinction between nature and man-made will be able to provide us with the legal guidance we are going to need over the next century with an explosion of new carbon- and noncarbon-based information-processing machines.

These existing patents apply to the final product of an altered genome, or to processes to alter it, but in some sense not to the genome itself. Genome patents raise a number of tricky technical questions in terms of the relatedness of genes and genomes. How distinct must genomes be, to be awarded two distinct patents? And what if a genome contains homologue genes, components that were patented separately by others, as will almost invariably be the case?[5] After all, the DNA sequences of human and ape genes are over 98% similar on average. The same issues can be raised for software. To demonstrate the complexity of these matters further, we need only to look at the new discipline of DNA computing, in which the chemical and storage properties of DNA molecules are used to store and process information, for instance, to solve combinatorial problems. What is the patentability status of DNA computers and DNA-computer software?[6] As exemplified by these problems, technology is getting ahead of public knowledge, and legal and ethical systems; and this of course brings an element of risk.

If gene patents are controversial, even more controversial patents have been awarded for transgenic animals. The first famous mammalian case was a patent awarded in 1988 to Harvard University for an oncomouse, a mouse with an oncogene, a gene capable of promoting cancer. This patent, bearing on the process rather than the genome, was initially commercialized by Du Pont and later by Charles River Laboratories in Wilmington, Massachusetts. You can order any one of a handful of slightly different oncomice for $60 to $90 per mouse, depending on the type of trans-oncogene.[7]

There are also patent issues regarding cells, organs, and basically any new biotechnology process or product. Literally thousands of patents have been filed to cover the properties of particular varieties of plants. And what about patents for new infectious agents that could be used, for instance, in biochemical warfare?

The rapidly evolving situation is also leading to strange legal strategies. Stuart Newman, a biologist at the New York Medical College in Valhalla, and Jeremy Rifkin, president of the Foundation on Economic Trends in Washington, D.C., filed a patent for a "humanzee," a human-chimpazee chimeric animal. These men have not created such a hybrid, nor do they wish to do so. The goal of their patent application is to prevent others from making a humanzee, or any other human-animal combination, and to challenge the current patenting laws. So far the Patent and Trademark Office has rejected the application.

Beyond these patenting questions looms in principle an even more daunting issue: What about patenting human genomes? It is clear that we are moving toward an era in which potentially the genome of each individual will be easy to sequence. We have seen, for instance, that most genetic differences between humans are confined to single-letter variations occurring at specific locations, on average one DNA letter every few hundred letters. Human DNA polymorphism will play an essential role in the medicine of the future, where treatments could be tailored to the genetic makeup of each individual. But will an individual "own" the rights to his genome? Should an individual have legal control over his genome and the dissemination of the information it contains? To what degree? And if not, who should? And if he does, what about his clones? Can there be genome co-ownership? For this and many other reasons, it is highly doubtful that human genomes will ever be patentable, and certainly not at any time in the near future.

On a related note, however, in January 2000 the California-based Geron Corporation received two British patents granting it commercial rights to human embryos created by cloning. A few months earlier, in the summer of 1999, Geron bought Roslin Bio-med, the commercial arm of the Roslin Institute, where Dolly was created. The patent covers only embryos at the very early stages of development and gives exclusive rights to "a reconstituted animal embryo prepared by transferring the nucleus of a quiescent diploid donor cell into a suitable recipient cell." One of the goals of Geron is to develop therapeutic cloning, whereby an individual is cloned by nuclear transfer, and after a few days stem cells are harvested from the embryo for treatment purposes. A similar claim, covering only cloning of nonhuman animals, has been filed in the United States.

### *Pragmatic Approaches*

A dogmatic absolute approach in ethical or legal matters, based on vague general principles, is clearly going to be impractical. We have to be much more flexible and also, because of the speed at which technology evolves, pragmatic and empirical. The philosophical foundation for such a type of approach was laid by utilitarian philosophers such as David Hume, Jeremy Bentham, and John Stuart Mill. Utilitarianism has several variants, but the central idea is that the aim of an action should be the greatest happiness for the greatest number. The determining consideration of right conduct should be the usefulness of its consequences. Thus the central formulation of utilitarianism is mathematical; it is an optimization problem. In principle it provides a clear direction, although difficult issues remain to be addressed because one must be able to define happiness and to carry out the optimizing procedure. Yet the inherently pragmatic and flexible nature of the approach is obvious: optimization must be done anew with each important decision, because the utility function will vary over time. In its most extreme twentieth-century mathematical form, utilitarianism leads to the framework of decision theory and game theory,[8] in which optimal decisions in uncertain environments are associated with the maximum of expected rewards.

Applications of decision theory and game theory, however, have so far been relatively limited for several reasons. Obtaining the happiness function, or all the costs, risks, and rewards associated with all possible actions and states of the environment, is certainly not easy when it comes to human matters. More fundamentally, perhaps, game theory is computationally intensive—only the most simple of formal games are solvable analytically. It is hoped, however, that with the increase in computing power such approaches may become more practical, and perhaps, in several decades or centuries, applicable to some of the difficult decisions that will have to be made, and to the modeling of complex emergent properties of social and economic organizations and systems. If you feel uncomfortable about optimizing cost functions when it comes to human life, as we all do, you ought to realize that this is commonly done by insurance companies and health care systems and providers. For now we can only dream about the collective utility or fitness functions that have to be optimized (Should each clone or each Siamese twin contribute as one?) and, farther down the road, how to compute averages over all information-processing systems in real time over the Internet. Weak forms of this idea are mushrooming and progressively will change communities, administrations, and governments. Indeed, censuses, opinion polls, and votes can be conducted in real time over the Internet. This is already common practice within smaller organizations, such as companies. In time, such procedures could progress toward increasingly more global levels of organizations to cities, states, nations, planets, and beyond.

In the meantime, as technology forces us to make difficult decisions, we can retain a pragmatic and flexible utilitarian framework in spite of difficulties encountered with the application of decision theory. We need a broad debate and flexible decisions that can be revised and updated as necessary, that take into account differences in beliefs and attitudes among all concerned entities at any given time, and that can be evaluated by trade-offs between risks and advantages they pose to our communities in the most rational ways, rather than by invoking absolute principles with dubious foundation and poor applicability. We can strive to progress prudently by taking small, incremental steps, even if we cannot always understand all their consequences or ensure that they are reversible, as long as enough benefits are derived for the community. Such small steps might help preserve our definition of human nature at a given time while permitting our adaptation to a revised definition at the next time step.

## Contemporary Issues: Cloning, Embryonic Cells, and Gene Therapy

*History tells us that not only is it unwise to try to hold back innovation, it is also not possible over the longer run.*

—Alan Greenspan, economist and chairman, Federal Reserve Board

*Cloning may be good and it may be bad. Probably it's a bit of both. The question must not be greeted with reflex hysteria but decided quietly, soberly and on its merits. We need less emotion and more thought.*

—Richard Dawkins, biologist and writer

### *Cloning*

Of all the scenarios we have discussed, human cloning is probably the most pressing and concrete. In fact, many countries, religious organizations, and international organizations have taken widely diverging positions on this topic[9] and, unlike in vitro baby technology, human cloning is essentially available today. In addition to some of the long-term issues addressed in this book, cloning raises obvious ethical questions from a standpoint of human nature and human dignity, not the least of which is the question of freedom and self-determination of a clone. Before human clones are produced, we should ask ourselves whether it is ethical for human beings to precisely determine the genome of another human being.

Even isolated cases of human cloning raise many legal questions that must be faced and carefully thought through. If nothing else, they set a precedent, and this alone is significant from the standpoint of jurisprudence. Who controls cloning? How do we decide whom we clone? Should there be any restrictions? Could we clone a Hitler?[10] Could and should one enter into contact with his or her clone? Should one be allowed to raise one's clone? Should we clone people who have died, or only people who are currently alive? Should we be allowed to clone our parents? Should sensitive information, such as medical, educational, or even criminal data, be shared with one's clone? Should we introduce a limit on the number of times a person can be cloned over a given period of time? Should organ transplants between clones be allowed? Many of these questions will also depend on the overall state of our technologies and societies at the time cloning is introduced, and how much we understand the relation between nature and nurture.

It is too early to provide definitive answers to such questions, but it is not too early to begin articulating them. From a utilitarian point of view, it is entirely appropriate and wise to have a short-term, revisable moratorium on cloning so that the issues can be better assessed. In essence this is the conclusion reached in the United States by the National Bioethics Advisory Commission in its June 1997 *Report and Recommendations to the President.*[11] In particular, we have to examine what is likely to happen and what are the risks and advantages of cloning. We need time to think through further and debate these issues, and the information must spread throughout organizations and societies. We have to make sure that clones do not age faster or do not show a higher susceptibility to genetic diseases, and this will require further experiments in animals. With current technology, for instance, cloned calves are larger at birth than normal newborns and have high blood pressure and difficulty breathing. But these characteristics seem to disappear entirely by two months of age. Large numbers of experiments must be carried out in order to be statistically certain that cloning, or any other artificial reproductive technique, does not carry an increased genetic risk of, say, 1/1000. Indeed, animal cloning is likely to continue unabated and to expand in the coming decades, for many purposes from biomedical research and applications, to agriculture, and even to the preservation of pets.[12]

Although we ought to be cautious, we should not harbor too many illusions. As Alan Greenspan said, in a slightly different context, not only is it unwise to block innovation, it is impossible over the long run. Thus history shows it is extremely likely that human cloning will happen sooner or later, if it has not already. It remains to be seen whether it will happen first in a legal or an illegal setting, with public or private funds, and in which country. At a deeper level, however, whatever its form and scale, it raises all the issues central to this book.

In my judgment, we do not have too much to fear about cloning in the short term, and we have plenty of time to think about its consequences if we begin now. It will take quite some time and debate before the first laws are passed authorizing human cloning, and it may take some time to achieve the level of technical proficiency required for its legal practice. It will take decades for the first human clone to become an adult, and for us to begin to sort out the effects of nature and nurture. The procedure will initially be isolated and expensive, involving an exceedingly small proportion of individuals and with no risk whatsoever of altering the human genetic pool or the composition of our population.[13] Initially, its diffusion is likely to be slow, which will provide us with additional time for reflection. We are at least several decades away from the time human cloning could be economically feasible on a large scale. After all, a well-understood technology such as IVF is still expensive and far from being widely available or completely efficient.

It is plausible that the first day human cloning becomes legal in one or a few countries, it will be legal only in a very restricted setting. It could in principle have numerous applications: from organ farming, to reproduction and partial immortality, to the study of nature versus nurture. Some of these applications, such as organ farming, are completely out of the question in the foreseeable future and, as we have seen, viable technological alternatives are under active development.

It will not be easy to justify human cloning to help infertile couples because many alternatives exist: adoption, IVF, surrogate motherhood, and so forth. Furthermore, even when both members of a couple are infertile, techniques are beginning to appear in which sexual reproduction can be simulated in vitro. This could be done by transferring the nuclei of the cells of the parents into two special cells that can go through mitotic division and produce two pairs of cells, each with half the number of chromosomes, then combining one cell from each pair to generate a new son or daughter genome.[14] The genetic makeup of this offspring would be determined randomly, as in sexual reproduction, by DNA recombination and chromosome assortments. This would alleviate one of the standard arguments against human cloning, whereby the cloned individual is born with less "freedom" both genetically and in terms of the potential psychological stigma and pressures that could be attached to it.

It is possible that cloning will first be legalized for a different application, vaguely reminiscent of animal pets in some sense, in the case of children who die by accident during the first years of life. For each newborn, the parents could ask that a few cells be stored as a repository of the infant's genetic information. In case of accidental death, say in the first three years of life, the parents would have a period of a few months over which they could decide whether or not to clone the dead child. This would avoid several of the drawbacks associated with cloning.[15] In particular it would have no danger of altering the human genetic population because one individual would be replaced by a genetically identical individual at most a few years younger than the original one. The process would aim at creating a life rather than destroying it and, for the parents, perhaps would alleviate the grief associated with the loss of a child and the attachment to its unique combination of genes and other characteristics. Furthermore, in general the clone would have the same biological and family parents as the original, and one could argue that because one is cloning an infant who still has a lot to learn, rather than adult, the issues of personal freedom are less affected.

Still, some will undoubtedly maintain that the cloned child could suffer psychologically by feeling like or being treated as a second-class citizen, a replacement of somebody else. Lack of genetic freedom, the fact that his fate is partially predetermined, and the associated projections and expectations from the parents could also place a heavy burden on his development. On careful examination, however, these arguments seem rather weak. Although we have no direct experience of such situations, it is clear that the determination would be strongest only up to age three years or so and progressively attenuate after that. Furthermore, throughout human history, there have been plenty of children whose birth and childhood were surrounded by very strong parental and even societal expectations. Queen Elizabeth II, for instance, knew from a young age that she would be queen of England one day. And what about the first child conceived in vitro? Finally, well-informed parents could try to minimize such effects—for instance, by not revealing the cloning operation until the child is able to understand, as is routinely done for adopted children.

The applications of cloning to immortality or some duplication of the self are either theoretical or distant in the future. In the near term, before it becomes part of everyday life or even obsolete, it is fair to say that human cloning seems to have relatively few useful medical applications. Therefore it is difficult to argue in favor of urgency in proceeding with its development. Rather, the controversy is concentrating around other precloning topics, such as human embryonic stem cells and gene modifications that have considerable scientific and medical potential.

### *Stem Cells*

Embryonic stem cells are derived from unwanted embryos or aborted fetuses. They can be grown in culture and are totipotent, or at least pluripotent. A pluripotent cell can, under the proper conditions, differentiate into a variety of other cells (humans have over 250 cell types) characteristic of different organs or tissues. The potential applications of embryonic stem cells are enormous and could greatly benefit biology and medicine. In the United States, the National Institutes of Health (NIH) and its former director, Harold Varmus, decided to allow funding of pluripotent stem cell research.

In response to this decision, in February 1999 seventy members of Congress, including Henry Hyde, chairman of the House Judiciary Committee, signed a letter calling upon the U.S. Department of Health and Human Services (DHHS) to reverse Varmus's decision and impose a ban on all research on stem cells obtained from human embryos or fetuses. The letter states, "Any NIH action to initiate funding of such research would violate both the letter and the spirit of the federal law banning federal support for research in which human embryos are harmed or destroyed." Indeed, right-to-life activists consider any destructive use of embryos immoral, and would like to see a total ban on embryo research, as in the eloquent statement of Jay Dickey, Republican representative from Arkansas: "We believe that science should serve humans, not that humans should serve science."

In March 1999, a letter supporting embryonic stem cell research signed by thirty-three Nobel laureates was sent to both President Clinton and Congress. A second letter, signed by a group of seventy-three scientists including sixty-seven Nobel laureates, appeared in the March 19, 1999, issue of *Science*. The scientists stated:

*In addition to helping to unravel processes underlying cell differentiation and biological development (which in turn, could lead to new ways to prevent and treat birth defects and cancer), the use of human pluripotent stem cells could potentially reduce the number of animal studies and clinical trials required for drug development and testing. The implications of this research for clinical medicine are equally enormous. Stem cells could be used to generate a long list of cells and tissues that could be used for transplantation. Myocardiocytes, for instance, could be injected into the heart, to heal myopathies and scars; neurons could be transplanted into the brains of patients with neurodegenerative disorders such as Parkinson's disease; and insulin-producing beta cells could be used to treat—or perhaps even cure—patients with diabetes. We join with other scientific organizations and patient groups in our belief that DHHS's current position is both laudable and forward-thinking. It succeeds in protecting the sanctity of human life without impeding biomedical research that could be profoundly important to the understanding and treatment of human disease.*[16]

I very much doubt that the concept of "sanctity" of human life is much more precise than the concept of "dignity." In fact, one can speculate that many of the signing scientists are probably not religious, and suspect that the religious connotation was used for other well-justified reasons. The pragmatic arguments in favor of stem cell research are certainly compelling. The letter continues: "[NIH has announced plans] to move forward in a careful and deliberate fashion to develop rigorous guidelines that address the special ethical, legal, and moral issues surrounding the research." Indeed, NIH proposed the creation of an outside committee to review grant proposals and make sure they are in line with any criteria set by Congress, main guidelines set by the 1994 Report of the Human Embryo Research

Panel, and regulations regarding research on transplantation of fetal tissues (section 498A of the Public Health Services Act).

In July 1999 the National Bioethics Advisory Commission recommended federal funding not only for research on human embryonic stem cells but also for the production of cell cultures, even at the cost of sacrificing embryos. The White House, however, adopted a more conservative legal stance: "No other legal actions are necessary at this time because it appears that human embryonic stem cells will be available from the private sector." And research on embryonic stem cells is "permissible under the current congressional ban." This position backs NIH analysis and interpretation of current laws so that government funds can be spent to study, but not to derive, stem cells from embryos; derivation is permissible only in private laboratories. This restriction does not apply to fetal tissue (table 10.1).

Embryonic stem cell research, like abortion and IVF, brings back the problems of the status of the embryo. In the United States, two sets of guidelines currently address the status of human embryos, both leaning against their use in research: the 1994 Report of the Human Embryo Research Panel and regulations regarding research on transplantation of fetal tissues (section 498A of the Public Health Services Act). The guidelines even prohibit the use of public funds for research on tissues derived from human embryos. Hence the need to distinguish legally among embryonic tissues and cells derived from them. Indeed, the first successful effort to derive human embryonic stem cells (Thomson et al. 1998) was privately funded by Geron Corporation.[17] In the United Kingdom, an age threshold of seven days has been set, before which the use of embryos in research is allowed. Whereas setting such an age threshold is probably important for practical matters, it also involves some level of arbitrariness and perhaps, in some cases, a drop of duplicity.

Not surprisingly, legal arguments have become quite subtle and rely on the distinction between fetuses and embryos, on the one hand, and cells derived from them, on the other hand. Other legal issues surround, for instance, the production of documents of origin and consent for the embryos. Donated embryos typically originate from clients of private fertility clinics that are not subject to the federal rules on informed consent. One key distinction here is that fetuses are usually derived from abortions, and

Table 10.1 NIH guidelines for research on embryonic stem cells as of December 1999

| | |
|---|---|
| Deriving new cell lines from embryos | Prohibited |
| Research on privately derived cell lines from embryos | Allowed |
| Deriving new cell lines from fetal tissue | Allowed |
| Research on cell lines from fetal tissue | Allowed |
| Research that would use stem cells to create a human embryo | Prohibited |
| Combining human stem cells with animal embryos | Prohibited |
| Use of stem cells for reproductive cloning | Prohibited |
| Research on stem cells derived from embryos created for research purposes | Prohibited |

Adapted from *Science,* December 10, 1999, 2050.
These NIH guidelines adopted in August 2000 further state that NIH-funded researchers can work on embryonic pluripotent stem cells derived by privately funded researchers provided that:

- Embryonic stem cell lines are derived only from frozen embryos created for fertility treatment;
- The decision to donate embryos is separated from fertility treatment;
- Embryo donors are told they cannot accept financial or other compensation and that the cells may be used indefinitely, possibly even for commercial purposes.

NIH funded researchers must also avoid using stem cells derived from nuclear transfer technology (cloning). In these revised guidelines, embryo donors can be identified provided they are notified in advance. Identifiers are essential to pass the Food and Drug Administration requirement for cell therapies aimed at minimizing immune responses and preventing transmission of diseases (adapted from the September 1, 2000 issue of *Science,* p. 1442).

are legally destined for death, whereas embryos are usually derived from fertility clinics, and are potentially destined for life. Yet there is a continuum from embryos to fetuses and, if anything, a fetus is closer to being human than an embryo is. Furthermore, what should private and public sectors have to do with the essence of human life? The current official position in the United States—without question a utilitarian one—may be the best compromise, given our history and the current state of our society. In my opinion, it also reflects our incoherent and irrational ways of thinking about biology and the essence of our species, and will have to evolve over time. In any case, this is clearly an area of legislation that is likely to undergo rapid changes in the future and across different countries.[18]

One point worth making is that current embryonic stem cell research is still generic. The goal is to study the general developmental properties of these cells and their applications, without choosing underlying genomes. Embryonic stem cells with a given genome are likely to lead to useful therapeutic applications to humans characterized by different genomes. In a number of situations, such as transplants or regeneration, however, it may turn out that the most suitable embryonic stem cells for a given individual are those with an identical genome. It is therefore conceivable that a few decades down the road we may derive and preserve these cells from each new embryo or fetus for possible future use. This is happening in some form today when blood cells from the umbilical cord are preserved for future uses. And this of course goes another step toward individual cloning. Embryonic stem cells can be used for reproductive cloning, but cloning of adult somatic cells can also be used to produce embryonic stem cells suitable for therapeutic and reproductive applications.

As I am about to finish this book, the U.S. Senate is debating whether embryonic stem cell research ought to receive public funding. There are an estimated 100,000 frozen embryos in the United States alone. One proposal trying to address most ethical concerns is to use only these already existing embryos, without creating new ones for the sake of research only, and provided there is consent from both parents and the embryos are not sold for profit. When Christopher Reeve and others with similar conditions are testifying, it is hard to believe that throwing these embryos in the garbage is more moral than trying to use them to save lives. Opponents of the proposal have suggested using adult stem cells only, which could have the additional advantage of their being genetically compatible and therefore being unlikely to be rejected. But in the present state of knowledge, the capabilities of adult stem cells are less promising for many therapeutic tasks. Because humans most likely can be cloned from many different kinds of adult cells and totitpotent cells can be derived from nonembryonic stem cells, the moral emphasis on embryonic cells and the debate about the use of embryonic versus adult stem cells will one day disappear. If a skin cell can be reprogrammed to become an embryonic cell, it is hard to argue that there is something sacred about embryonic cells. Ultimately, either we manipulate cells as large molecular assemblies or we don't. If we do, then the origin of the cells will make little difference to an advanced biotechnological society.[19]

## *Genetic Manipulation of Humans*

Another area of biotechnology loaded with ethical and legislative issues is gene screening, gene customization, and gene "fixing" or enhancement. Here, too, based on general absolute principles, several people and organizations have advocated a total ban on any artificial modification of the human genome. In this ban, the dignity of human life and its freedom are inseparable from a genome that should not be controlled in any way and whose composition must be forever left to chance events. From this point of view, it is difficult and potentially very dangerous to begin to define what a "defective" gene is. And customizing and selecting genomes could lead to abuses and catastrophes.

From a more utilitarian point of view, however, this position is rather difficult to sustain. A small but nontrivial percentage of the human population suffers greatly from simple genetic accidents, such as an unfavorable mutation in a single gene. This is the case, for instance, with sickle-cell anemia, cystic fibrosis, insulin-dependent diabetes, and Duchenne muscular dystrophy.[20] It is obvious that such individuals would have benefited greatly if the corresponding gene could have been corrected at the time of their conception. I doubt that any one of them would have refused such an opportunity if given the choice. This kind of gene therapy is useful, reduces total human suffering, and hence is ethical.

There is a list of ailments for which the notion "defective gene" is completely unquestionable, and we can start to work on these. Gene therapies for the treatment of several genetic diseases in adults and children are under active investigation in many medical centers. Some genetic diseases cause irreversible damage to the fetus before birth or would benefit from early intervention. In utero gene therapy is technically conceivable and, in fact, is under active investigation in animals. It obviously raises additional ethical questions, and further work is required before clinical trials in humans and the associated risks can be addressed.

Of course, we must carefully regulate progress in these areas and think things through to avoid abuses or completely unforeseen and unwanted consequences, such as the well-publicized death of a volunteer patient in a gene therapy experiment at the end of 1999. We have time to do exactly that before we consider moving toward more difficult issues, such as multigenic diseases or complex traits, such as body size or intelligence. The precise course of events and the corresponding legal and ethical challenges over the next few decades, however, will be intimately interdependent with technological progress and its unpredictable breakthroughs.[21]

There are many other important issues in bioethics, a fair discussion of which would require a book of its own. But another problem worth mentioning briefly is privacy. How private should genetic information be? As we have seen, the technology for imaging the level of expression of our genes in any one of our cells is here and, cost issues aside, we are capable of sequencing entire human genomes. Potentially this could be extremely useful for developing diagnostic and therapeutic techniques tailored to each individual, even perhaps without any side effects. The process of finding individual treatments, however, inevitably requires maintaining large databases of genetic information, and this in turns creates many privacy issues. For instance, making genetic information public or widely available increases the risk of discrimination by employers, insurance providers, and so forth. Again this is a dynamic problem that will require different approaches and compromises at different times in history, depending on the state of our societies and technology. In current practice, study subjects can decide whether their individual information will be identified or not, even though all study samples are identifiable so that clinical data can be linked. After the study, samples are destroyed or stripped of identifiers, and banked indefinitely. The National Bioethics Advisory Committee has suggested that DNA ought to be traceable back to each individual in case research on the same disease is pursued in the future.

Cryptography is likely to play a role as well, one more area of intersection between information and computer sciences and biology. Perhaps one day each person will have his or her genome on his or her web page. Perhaps it will be cryptographically protected so that portions of it are available only to certain classes of users. Perhaps, for a while, this will be part of the ultimate strategy for personalized medicine and matchmaking.

Kant used to marvel at the stars above us and the ethical values inside us. In a much more science fiction vein, we could finally wonder what would happen if silicon chips could be implanted in our brains that modify our sense of justice and of ethics, and our ability to feel guilt and right. What should be the legal rights of a silicon copy of a human brain? Luckily, we do not have to worry too much about such issues for now.

## What Can Go Wrong

*All evil comes from ignorance.*

—Confucius, Chinese philosopher (551–479 B.C.)

*. . . education is an essential human right, a force for social change—and the single most vital element in combating poverty, empowering women, safeguarding children from exploitative and hazardous labor and sexual exploitation, promoting human rights and democracy, protecting the environment, and controlling population growth. Education is a path toward international peace and security. . . .*

*One hundred-thirty million children in the developing world are denied this right—almost two-thirds of them girls. Nearly one billion people are illiterate—the majority of them women.*

—Kofi Annan, secretary-general of the United Nations, in *Forward, the State of the World's Children* (1999)

It is not too difficult to come up with a number of catastrophic scenarios that could halt human technological progress, or the human race altogether. According to one of the dominant theories, dinosaurs were wiped from the surface of Earth by the impact of a large meteor roughly 60 million years ago. During the cold war, nuclear disaster was perhaps prominent in people's minds. The greenhouse effect and the AIDS epidemic remind us that environmental collapses and new epidemics are also possible. Biotechnology could bring dangers of its own—for instance, via the release of new organisms in the environment.

Bioengineered organisms raise more complex problems than bioengineered products because they can be more difficult to control. A classic example is plants that are genetically engineered to resist herbicides. Herbicides are commonly applied to crops to kill weeds and other undesirable plants that may compete with a desirable species. However, they can also be harmful to the desirable species. Genes that confer resistance to common herbicides, such as glyphosate, have been identified and inserted into several different plants. Larger amounts of herbicides can be used on such genetically engineered crops. But is that entirely a good thing? And what would happen if some of the herbicide-resistant genes were laterally transferred—for instance, by a plant virus—from the desirable plants to the undesirable ones? There are a number of cases, however, in which the advantages of bioengineered organisms seem overwhelming, and we may concentrate on those ones first. For instance, one of the most valuable genetic operations would be to create corn or wheat crops that can fix nitrogen directly from the atmosphere. Indeed, several bacteria exist that are capable of capturing nitrogen from the atmosphere. If such genes for nitrogen fixation could be successfully inserted into corn or wheat plants, the use of nitrogen fertilizers could be substantially reduced. This would greatly reduce the costs and pollution associated with such fertilizers.

Some of the extreme dangers are obvious, and it is unlikely we will take any step in their direction in the foreseeable future, even if the technology were available to us. In the next few decades, for instance, it is highly unlikely that we will use artificial means to abruptly manipulate the human gene pool or the composition of the human population in a massive way. It is also unlikely that we would decide to make humans smaller just so fewer resources would be consumed. Other dangers, however, might be less visible or become noticeable to us only after a certain time. We could cross important irreversible thresholds without being aware of it, and produce disequilibria that are hard to fix retroactively. We are particularly vulnerable during the transition phase we are entering, with its mixed regime, where natural and artificial evolution coexist, where the old principles do not apply entirely but the new ones are not completely understood. Astronomical catastrophes are relatively rare and occur on timescales of millions of years or longer, but the revolution discussed in this book is occurring on a much faster timescale—a few centuries at most. Human cloning is already knocking at our doors.

In the list of dangers that are perhaps a little more subtle, could our brains become destructively addicted to newly engineered drugs? Or to new computer games and virtual realities? After all, the game industry has easily surpassed the Hollywood film industry. Could an insurmountable collective existential crisis result from the discrepancy between who we are and who we think we are, between the view of the world slowly engraved by evolution in our brains over eons and the new view rapidly emerging from our technologies? Are we wise enough to control or direct our own evolution? And how small a risk is acceptable?

Many catastrophic scenarios have been explored in the science fiction literature, and any serious discussion of them would require a book of its own. All we can say is that so far, humans have shown a remarkable degree of resilience, adaptability, and ingenuity that has enabled them to survive and adapt to many crises and changes throughout their history. After all, brains did not evolve in a world full of cars.

The adaptability of the human brain is indeed a very difficult question. Physicists and chemists study what happens to matter and to different compounds when they are subjected to extreme conditions of temperature, pressure, magnetic energy, and so forth. Such conditions are outside the realm of our experience and in some instances even outside what the universe may have known since the Big Bang. In a sense, the same is happening to the human brain—probably the most complex system we know in the universe. It is progressively subjected to entirely new situations that were not present in the environment that served for its construction. Things like computers, the Internet, cloning, and a form of sexuality that is irrelevant to reproduction. We do not know at all how a brain reacts when it is raised in the presence of twenty twin brothers or in a world where sex is reproductively futile. Furthermore, these new environments are occurring at an accelerated pace that seem to far exceed the timescale at which natural evolution can adapt the brain. As if this were not enough, it is already unclear how natural evolution still operates on humans these days. And the brain is now beginning to manipulate genomes in ways that will render human natural evolution obsolete.

Consider the case of a human raised with a large number of clones. Twin studies have claimed that identical twins raised together tend to be more different than identical twins raised apart. Though this is something difficult to establish with scientific precision, it could result from internal differentiation mechanisms present in the brain. Such mechanisms work fine in the case of twins. But could they break down in the case of 20, 50, or 1000 clones, where the task of differentiation becomes a nightmare? Most of us do not wish to attempt such experiments for now, but it is not unthinkable that such a threshold exists. Perhaps when a large number of identical twins are raised together, extreme conformity, rather than over-differentiation, sets in. We just do not know.

In a related but more general vein, forces toward conformity and differentiation coexist in our brains and interact in complex ways with our societies. One can be concerned about the general trend toward uniformity—one McDonald's per square mile—that is driven by the globalization of our economies, the media, and especially the Internet. Could such a trend lead toward excessive conformism, cultural homogeneity, and a lack of diversity and creativity—or toward the opposite, a community of freaks desperate to exploit any random fluctuation to assert their diversity? And how dangerous is that? After all, a world dominated by one of these extreme perspectives appears depressive only from our current self-centered and individualistic viewpoint, and this may also change. It is impossible to predict whether over time these opposite forces will coexist and remain more or less in balance, or whether one will begin to dominate the other. Technology alone, including the Internet, provides new means for both standardization and differentiation.

A second area of concern for the medium term is the area of sex and relationships between men and women. The danger is heightened by the fact that in the medium term, sex affects a substantial fraction of our brains and behaviors, as well as the composition of the human population. Furthermore, this is precisely an area where things could change very fast as a result of developments in both biotechnology and computers/Internet. There exist biological and behavioral differences between the sexes that are connected to our evolution and that are beginning to be recognized. In spite of the old romantic ideal, it has become clear that such differences do not always work in the direction of greater harmony within couples, but can be associated with competition and tensions. After all, evolution cares more about reproduction and survival than about marital harmony—high divorce rates should not come as a surprise. But we have seen that sex is on the verge of becoming completely obsolete as a reproductive function and losing its original raison d'être. Yet, is it that easy to adapt to a sex for love and pleasure only? Some may argue that such adaptation is easy, and has already occurred with the invention of the birth control pill. I am skeptical that it is so simple, because the quality of the present changes is entirely different. With contraception, the bedroom became *insufficient*. With assisted reproductive technologies, it is becoming *unnecessary*. Contraception slightly modifies one function. Modern biotechnology massively modifies our essence. Past the evolutionary bootstrap, sex becomes functionally meaningless.

This raises two disturbing alternatives: either sex (both the activity and the gender) as we know it will progressively disappear, or it will have to be reinvented and evolve into something else. In both cases, this should substantially change who we are. Actually the alternatives are not exclusive, and signs of both trends can already be detected. But how will the male and female brains cope with that? It is a cliché to describe male sexuality as being less emotional and more visual and factual. This is probably one of the main reasons behind the popularity of different forms of cybersex over the Internet and the fact that Internet pornography has been at the forefront of E-commerce. Although computers and the Internet have the potential for expanding our sexual experience, this phenomenon should not be taken lightly, and has already been reported as a cause of withdrawal from traditional sex and of stress in couples. Together with the progressive loss of their reproductive prerogatives, this is a trend that should worry women first[22] (but also men), and one in which a broader debate is perhaps needed. Sex started as a biological-to-biological affair, one of the main driving forces behind life on Earth during most of its history. Its biological-to-digital version is now emerging. Will its digital-to-digital version dominate in a distant future of information-processing orgies over the Internet? And if so, will it be more than mere communication?

In many respects, our greatest enemy is probably ignorance. We have begun to tinker with complex, interacting dynamical systems and equilibria, such as the human brain and the environment, that we are far from understanding completely, let alone controlling completely. Suddenly we have the means to introduce major perturbations in such systems without the time and luxury of understanding all the consequences. They are major perturbations by their size and their speed, but also because they were never encountered by the systems during the millions or billions years over which they were being laboriously constructed by evolution. Ecosystems and brains were built slowly, according to very ancient principles. At present we do not have all the information and tools to predict the future behavior of such systems because of their sheer complexity, and perhaps because of other more fundamental limitations, such as chaotic behavior.[23] In my view, these are all good arguments for prudence but also for more science, more understanding, rather than less.

Science has been one of the greatest forces driving the evolution of human societies over the centuries. In spite of a few hiccups along the way, the majority of its effects have been, overall, positive. Human quality of life, health, and life span have greatly improved over the last few hundred years thanks to science. Not only has science given us the technology to lead longer and better lives, it has also given us the fundamental knowledge by which we can understand and place ourselves in the universe. Clearly, there are still plenty of problems to be addressed—eradicating famine, curing more diseases, protecting the environment, reducing the gap of wealth between countries and the gap of literacy between people,[24] to name just a few. But there is no reason why, over the long haul, science could not be one of the main driving forces behind their solution, even if in the short term some of these problems may worsen. It has been said that in a sense science is neutral—it can be used or misused by humans. This is true of scientific knowledge, but not of science as a human and societal enterprise. It is a mistake for scientists to believe that the use or misuse of science does not depend on them, but only on politicians. Scientists should become more involved—not less—with promoting science and helping guide human technology toward human betterment.

Figure 11.1 Like many twin pairs, the models for David Teplica's photograph *Ovum, 1997* lead an entangled adult life. Upon arrival at the studio for their photographic session, the young woman asked "We came from one egg and then split into two *separate* people, didn't we?" [Both the question and the emphasis were theirs.] The image visually captures the twins in a way that explores their incomplete individuation and meshed psychologies. (Photograph copyright by David Teplica.)

# The Information Space

# 11

*But there is nothing magic about Darwinian fitness in the genetic sense. There is no law giving it priority as the fundamental quantity that is maximized. Fitness is just a way of talking about the survival of replicators, in this case genetic replicators.*
—Richard Dawkins, biologist and writer, in *The Extended Phenotype*

## Biology.com

Next time you walk into a crowd, say in a shopping mall, try to imagine that you are surrounded by "gene machines." For me, it works best in the Irvine Spectrum shopping center. Irvine is a futuristic city built in the southern California coastal desert, halfway along the metropolis that now extends almost continuously from Los Angeles to San Diego. Today, it is the largest master planned urban environment in the United States. It is being developed by the Irvine Company, following a comprehensive master plan created in the 1960s. This makes Irvine an intriguing urban environment consisting of several villages. In each village, for instance, banks are difficult to rob because they are systematically surrounded by an irregular maze of streets that does not facilitate quick escape toward the freeways. Bike lanes are ubiquitous, but seldom used, and seem to serve the rather different purpose of disallowing street parking. Within Irvine, Irvine Spectrum is the ultramodern local Silicon Valley ("Southern California Technology Coast"), mushrooming with thousands of high-tech established companies and start-ups in the middle of strawberry fields and other oddities of the California desert. If you want to get an eerie feel for what a biology.com society of the future could feel like, visit the Irvine Spectrum shopping center.

The Irvine Spectrum mall is a compact, perhaps the size of four blocks, oasis in the middle of a large parking lot, in the middle of the southern California desert. Indeed, it is built around the theme of an Arabian oasis. Both the compactness and the Arabian theme accentuate its strange and eerie feeling. At one of the two main entrances is a large Casbah, a Dave and Buster entertainment and gambling restaurant, with a gigantic game arcade. Throughout the mall, even the pay phones are framed by small Arabian-looking booths. It is like a large movie set, except it is not made of cardboard; an odd cube of craziness where reality and fiction seem to blend seamlessly. Strolling human aliens of all ages appear to be in a state of permanent vacation, with a strange mechanical gait, aimlessly floating through a bazaar of mostly gadget stores with names such as Limbo Lounge and O'My Sole.[1]

In this book, we have tried to look at some of the possibilities raised by modern computer and biological technologies. These possibilities raise questions about the nature of the world and the human condition, regardless of whether or not they are ethical, costly, or even likely to happen. Predicting the future is important for decision-making and setting policies. Yet history shows that most of our predictions are wrong, at least in their fine details. Jules Verne,[2] or even Leonardo da Vinci, two masters of fiction science, anticipated some of the technology of the twentieth century, from submarines to trips to the moon. Jules Verne accurately predicted in the late nineteenth century that man would set foot on the moon one day. He wrongly guessed that it would be achieved through the use of a large gun rather than a rocket. Gun or rocket is absolutely essential for the actual travel to the moon, but it makes little difference to the concept or possibility of a man on the moon.

Whereas the extrapolations in this book are grounded in common scientific knowledge, undoubtedly I have oversimplified several issues and systematically resorted to first-order approximations, approaching multifaceted problems along just a few dimensions. But this is a place to start rather than end.

One of the themes of this book is that we are witnessing a revolution brought about by scientific and technological advances, one in which change occurs at an accelerated pace. By 1945 we had the ability to destroy life on a large scale. By 2045, only a hundred years later, we ought to be able to create life from scratch, both in real space and in cyberspace. But why call it a revolution, and why is there acceleration? After all, science and technology have been progressing rapidly for several centuries. There are several answers. First, in the past centuries a lot of progress took place in the basic sciences: mathematics, physics, chemistry. Only since about 1950 has the pace tangibly accelerated in biology and computer science, and these are the sciences that study what we are made of and how we can somehow replicate ourselves. Second, we are in the process of passing some key thresholds, such as sequencing the human genome and developing computers with brainlike processing speeds. The third point has to do with the peculiar taxonomic nature of biological knowledge. Biology is very complex, but it has an inherent, discrete, and finite nature at the molecular level: the human genome is 3 billion letters long, with about 40,000 genes, and the universe of proteins can be classified into a few thousand families. The classifications are complex but finite, and can be handled by current computers. Sooner rather than later, we will know all the letters and genes in the human genome, and all the protein families, as well as their structures and functions. In each case, because the number of classes is finite, the classification process tends to accelerate toward its end. And we are now moving toward the more challenging problem of understanding all protein networks and metabolic pathways. Again a complex but inherently finite task. From there the applications seem limitless, with medicine on the verge of becoming a science. Finally the Internet, with its asymptotic form of instantaneous universal communication, is greatly accelerating the pace and volume of information propagation.

In many ways, we are reaching the end of our evolutionary odyssey. As in a simulation of artificial life played in the material world, natural evolution has been building increasingly sophisticated information-processing systems up to the current threshold, where the most complex systems are getting ready to take on a life of their own. All the things that have been created and molded by evolution stand a chance of being seriously challenged.

I have used the word "end"—the end of natural evolution—because it is both simple and concise. It should be clear, however, that natural evolution is nowhere near its end. Its laws continue to apply unabated, from the depth of the oceans to the jungles of Brazil, to the laboratories and cities of planet Earth, to life on other planets. The "eclipse" of natural evolution perhaps would have been more accurate, but obviously more obscure. Natural evolution is there, but "unnatural" evolution is casting its shadow. As for the word "unnatural," it is also a form of speech; after all, unnatural evolution is the product of humankind, which is a product of nature and natural evolution. Unnatural evolution still obeys the law of nature and in many ways extends natural evolution. There ought to be a broader concept of evolution instead, one that would include unnatural evolution and the evolution of information systems, as well as natural evolution as a very primitive special case. It is with respect to this broader evolution that natural evolution in the end seems artificially restricted.

The irony and paradox is that by necessity, natural evolution had to operate in an artificial world, the tip of an iceberg with no computers, human cloning, or Internet. Our brain has evolved in this information-poor artificial environment for millions of years, with little knowledge of the rest of the iceberg—which is now rapidly becoming apparent, with its endless possibilities for information storage and transmission, and virtual continuum of genomes and phenotypes.

Biotechnologies and the Internet are placing the human brain in the real natural world, the hidden part of the iceberg, totally different from the artificial world of natural evolution that was used to manufacture it. In this new reality of more or less continuous genotypes and phenotypes, all kinds of new creatures are beginning to pop up at a rapid pace, forcing us to revise our concepts, our laws, and our sense of whatever makes us human. Genomes can be manipulated just like software, and genetically identical complex organisms can be mass-produced like the Ford Model T, without a single drop of sex. External selves can be recorded and posted on the Internet, forever.[3] We have just scratched the surface of what can go on in our minds. What kind of bizarre event, of rare alignment of neuronal stars, is behind Oliver Sacks's twins, Tourette syndrome, and all the other fantastic aberrations of the human mind described in the literature?

It is a little as if all along the human brain had modeled the world, using visual illusions, and incorporating these illusions into its wiring in complex ways that we are just faintly beginning to understand. We have been wired and raised to have the inner feeling that each of us is "one," a well-defined and distinct unit, monolithically isolated and different from other beings, with a precise physical start and end. In fact, the genome space and the information-processing space are almost continuous and densely populated, and only rare isolated samples have been available to us so far. Psychologically we have become attached to these isolated samples, and in particular to our own. In reality, we are much more diluted and continuous both in space and in time, arbitrarily close and connected to many beings, most of which are virtual but could become material at any time.

Because of our technological advances, these illusions are now waning, leaving us at a place of mental discomfort that is inversely proportional to the material comfort generated by the technologies. One of the keys to human successes is the ability to construct increasingly complex models of the world and align them with reality. It is this alignment that is now being challenged and requires a revision of the basic model. What it should be replaced with, however, is much less clear, in part because of our limited knowledge of the brain and of large computational systems. We do not know who we are, but we know enough to know that we are not who we think we are. Who we think we are is the result of an illusion produced by our evolutionary odyssey. As we begin to enter the realm of very large computational systems, a thick fog sets in. Qualia and the psychology of large numbers, strangely far from our daily experience but at the same time close to our technological limits, begin to take their toll. We do not yet fully understand the laws of computation for such large systems.

We have estimated the sizes of the external and of the internal self. In the course of extrapolating, somewhere between $10^{20}$ and $10^{30}$, the large numbers are getting the upper hand. As we move into the uncharted waters of $10^{30}$-plus computations per second and recorded bits, the fog thickens, we lose our way, and fiction science becomes science fiction. We can clone humans, exceed the computational power of current brains, record and store external selves, and connect any two information-processing devices. But when we move toward the internal self, problems become more nebulous. By manipulating genes and the environment, it is likely we can produce very similar brains, but how similar is unclear. I doubt that similarity at the level of molecules, or even single neurons, is possible, or that real-time, detailed simulation of all the computations a brain performs is possible. Perhaps there are fundamental limitations in this direction, something like "the number of particles in the universe is not sufficient," but we just do not know at this stage. Common words begin to lose their meaning. We can say that two brains or a brain and a machine are 80% identical. But what does it really mean? What if a man and a machine were to provide the exact same answer to 1 billion binary questions? It is both exciting and unsettling to know we are going to have the answers soon.

## The Information Space

*. . . he deduced that the Library is total and that its shelves contain all the possible combinations of the twenty-odd orthographic symbols (whose number, though vast, is not infinite); that is, everything which can be expressed, in all languages. Everything is there: the minute history of the future, the autobiographies of the archangels, the faithful catalogue of the Library, thousands and thousand of false catalogues, a demonstration of the fallacy of these catalogues, a demonstration of the fallacy of the true catalogue, the Gnostic gospel of Basilides, the commentary on this gospel, the commentary on the commentary of this gospel, the veridical account of your death, a version of each book in all languages, the interpolation of every book in all books.*

—Jorge Luis Borges, writer, in *Ficciones,* "The Library of Babel"

An old question that goes all the way back at least to the Greek philosophers has to do with the nature and reality of mathematical objects. In particular, do mathematical objects exist independent of us, of our thinking, and of their material incarnations? Consider a circle. It is hard to conceive that circles could not exist without human beings. And it is also hard to see why the notion of a circle and its mathematical properties, such as the formula for the length of its perimeter or for its area, should depend on any physical embodiments in our material world. Plato postulated that mathematical objects belonged to a special class of objects and formed a special world of what he called ideals.

If we are ready to think of a circle as a special type of object living in a special space, then it is clear that any integer—including 0 and 1—should be living in the same space and for the same reasons. It does not depend on us and or on a specific material realization. And if we accept the latter, then any string of zeros and ones should be living in that same space—let us call it the information space. It should be clear by now that the information space is crowded and contains, in particular, all possible genomes. We just have to code A = 00, C = 10, G = 01, and T = 11 to see that any DNA string can immediately be converted into a binary string, and any binary string of even length can be converted into a string over the ACGT alphabet. For the same reasons, this space contains all possible internal and external selves.

Today is June 26, 2000. Celera Genomics and the Human Genome Project Consortium just announced the completion of the first working draft of the human genome. As the bulk of the iceberg continues to emerge, the information space will become increasingly dominant. We do not completely understand it, and it is difficult, perhaps even meaningless, for us to completely ignore the material realizations of the objects that inhabit the information space. But as the relationship between the information space and the material space becomes more and more flexible, the fruits of natural evolution begin to lose their traditional meaning. We return to the small beach where we started: the boundary where marine life progressively evolved into land animals and plants, where the sand grains are as numerous as the neurons in our brains, and from which silicon-based information-processing computers and optic fibers emerged. And when atoms are scaled to sand, tiny sandstorms represent neuronal electrical activity. Like footprints on the beach, the self may progressively disappear in the ocean of the information space—its relic will become our Internet address on the information superhighway.

# Appendix A
# The Total Number of Humans

To a first-degree approximation, let us first assume that the population has been growing geometrically with a ratio of 2. This means we assume that, on average, each human has two distinct descendants in the next generation, or equivalently that each couple has four children who survive long enough to reproduce themselves. A simple calculation then shows that the total number of humans, past and present, is twice as many as the current number. This provides an estimate of about 12 billion (table A.1). This is, of course, only a very rough order of magnitude because population evolution is much more complicated. Birthrates and death rates, as well as life spans, have varied a lot throughout our history. Also, if we were to assume a geometric progression with a ratio larger than 2, the total number would be less than 12 billion. Table A.1 gives the number of generations necessary to reach a population of 6 billion for different ratios, as well as the cumulative number of individuals. A ratio of 2 yields thirty-three generations. So if a population begins with a single individual and doubles at each generation, the total number of individuals after thirty-three generations is roughly 6 billion. If we use an estimate of thirty years for the time span of a generation—a generous estimate for early humans—this provides a rough horizon of 1000 years for the duration of human history. This is a gross underestimation resulting from approximations made in the crude geometric model.

Note that the much larger number of all possible humans at the genetic level—all possible human genomes—is also finite. In fact, this is true also of all possible genomes, across all organisms that satisfy the current rules of

Table A.1 Numbers of generations and cumulative numbers of humans (in billions) with crude geometric model of population growth

| Ratio | Number of generations | Cumulative number of humans |
|---|---|---|
| 1.1 | 237 | 66.0 |
| 1.2 | 124 | 36.0 |
| 1.3 | 86 | 26.0 |
| 1.4 | 67 | 21.0 |
| 1.5 | 56 | 18.0 |
| 2.0 | 33 | 12.0 |
| 3.0 | 21 | 9.0 |
| 10.0 | 10 | 6.6 |

A ratio of 2 indicates that from one generation to the next the total population doubles. The number of generations is calculated so that the current population is 6 billion, roughly the size of the human population in the year 2000. A size increase of 10% per generation corresponds to a 1.1 ratio. With 237 generations, each 30 years in duration, this model yields a modern human history scale of about 7000 years. This is again an underestimate. It is not completely off, however, because on an evolutionary time scale the definition of human becomes fuzzy. It is not obvious, and is even unlikely, that modern human populations could interbreed by natural means with their hominid ancestors 1 million years ago, or even 100,000 years ago, assuming these were available. For comparison, humans cannot interbreed with gorillas or chimpanzees. Although all three have relatively similar genomes, they are packaged slightly differently: the number of chromosomes is not the same. Chimpanzees, for instance, have forty-eight chromosomes, compared with forty-six in humans. From a purely technical standpoint, it is conceivable and even likely, however, that a man-ape hybrid could be created in the laboratory using modern biotechnologies. Mice with extra human chromosomes have already been created.

life on Earth. To see this crudely, it is enough to derive an upper bound L on the maximum length of DNA contained in a cell—for instance by considering the maximum size and weight of a cell and the size and weight of DNA. Without considering a partitioning of DNA into chromosomes that would add a finite factor of complexity, the total number of possible—not necessarily viable—genomes is bounded by $4^L$, because there are four possible letters at each of the L positions.

# Appendix B
# Historical and Technological Aspects of Cloning

To be clear, it is important to understand that there are many different forms of cloning with many decades of research behind them. In the first form, cells from a given tissue or bacteria are allowed to divide in a petri dish culture. This produces a population of genetically identical cells that is useful for many biological tasks but falls short of ensuring the genetic duplication of a complex multicellular organism.

The second kind of cloning is simple nuclear transfer. Without going into many important experimental details, this consists basically of removing the nucleus of a mature egg (more precisely, an oocyte) and replacing it with the nucleus of an early embryonic cell, and starting the cell division process. This type of cloning is a direct ancestor of the technique used for Dolly. It was first realized in the 1930s in simple unicellular organisms, such as amoebas, under the influence of the German biologist Hans Spemann, 1936 winner of the Nobel Prize in medicine (he unfortunately ended up embracing the Nazi movement during World War II). In the 1950s such techniques were extended to frogs (*Rana pipiens*) by Robert Briggs and Thomas King in Philadelphia, and further refined on other amphibians (*Xenopus laevis*) by John Gurdon at Cambridge in the 1960s. Illmensee did some work around the same time on fruit flies (*Drosophila melanogaster*). Not until the 1980s, however, were successful nuclear transplant cloning experiments carried out on mammals: on mice in 1981 (Illmensee and Hoppe), sheep in 1986 (Willadsen), bovines in 1987 (Prather et al.), pigs in 1989 (Prather et al.), goats, and even rabbits in 1990 in France by the team of Dr. Renard.

It is important to realize that the goal of these early nuclear transfer experiments was to address a number of fundamental scientific questions about development. In particular, the question of the relationship between cytoplasm and nucleus during development and the question of potency—which cells are capable to develop into a full-blown organism (totipotency) and why. How does the genetic potential of a nucleus vary during embryonic development and differentiation? These early experiments showed that it was important to use the cytoplasm of an oocyte, but that the cytoplasm of a fertilized egg was not needed. They also showed that early embryonic cells are totipotent, and the experiments become increasingly difficult as the "developmental distance" between the donor nucleus and the receptor cytoplasm increases. They led to considerable improvements in the experimental protocols, such as the use of micropipettes, of a resting or even cooling phase for the donor cells or the newly implanted egg, of frozen embryos, of solutions that can soften membranes (McGrath and Solter 1983), and even of small electric fields to induce cell fusion (Zimmerman and Vienken 1982). They clarified a number of questions and differences between species.

Early embryonic development typically goes through four stages: morula, blastula, gastrula, and neurula. The morula corresponds to the first four to five days, with four to five division cycles and sixteen to thirty-two cells in a more or less spherical aggregate (*morula* means "mulberry" in Latin). The blastula extends up to seven to eight days, when the embryo begins to assume a shape faintly reminiscent of the final organism. The three primary germ layers of an embryo (ectoderm, mesoderm, and endoderm) appear during the gastrula phase. Nerve tissue begins to differentiate, and the basic pattern of the vertebrate starts to emerge during the neurula. During development, even the basic body plan is not specified all at once, but through a series of hierarchical decisions. A brain neuron can come only from the neural plate of the neurula, which can come only from the ectoderm layer in the gastrula.

The first successful nuclear transfer experiments used the nuclei of blastomers corresponding to cells in the morula or blastula phase of early development. These experiments also showed that one fundamental problem to be addressed in cloning is synchronization of the acceptor cytoplasm with the donor nucleus. During embryonic development, the first cell must go through a rapid series of divisions. At each division the nucleus and cytoplasm must communicate. Proteins produced in the cytoplasm must go back to the nucleus to initiate DNA replication and gene expression. The mRNA of genes that are transcribed in the nucleus must go to ribosomes in cytoplasm to be translated into proteins. Thus the cytoplasm and nucleus must dance in step with respect to DNA replication and cell division. If they are not in phase, the cloning operation fails. Putting the cells to rest, starving them, giving them particular nutrients, and cooling them are some of the techniques used to reset cellular programs and achieve synchronization. Furthermore, the first cell divisions are under the control of the cytoplasm. At some point there is a transition of control from the cytoplasm to the nucleus when the genes in the nucleus start to be massively expressed. This transition occurs at different times in different species: 1 to 2 cells in mice, 4 cells in pigs, 8 to 16 in sheep and cows, and 3000 to 4000 cells in amphibians. This in part explains why the first successful results were obtained on frogs. In humans the transition occurs between 2 and 4 cells.

The third form of cloning is artificial twinning. Identical twins are the result of an accidental split of the early embryo into two genetically identical embryos. Because during the first cycles of cell division the cells remain relatively undifferentiated, it is clear that this splitting event can be reproduced artificially. In principle, it could be repeated several times to yield arbitrarily many clones of a given organism. As early as 1892, Hans Driesch showed that each cell of a two-cell or four-cell sea urchin embryo is capable of developing into a separate, completely formed embryo. Artificial twinning in higher organisms was first done in the late 1970s by Steen Willadsen in sheep, and later extended to cows. Recently it has been achieved in primates. In primates, the technique is still not very efficient. In an experiment on rhesus monkeys reported in January 2000, 368 embryos were created by splitting 107 embryos into two or four pieces. Four pregnancies were obtained in thirteen tries but only one survived, resulting in a healthy female named Tetra (for the fact that she came from one-fourth of an embryo). With artificial twinning, one of the embryos can be frozen and harvested at a later date for stem cells that could be used to test various therapies, including organ regeneration and transplant, on the adult version. A most notable fact is that the initial splitting operation was even carried out on nonviable human embryos by Jerry Hall and Robert Stillman of George Washington University in 1993. Artificial twinning, however, is not true cloning because the starting point is always a set of early embryonic cells, up to about six days old (morula or blastula stage).

Full-blown cloning, on the other hand, consists in taking as the starting point almost any cell in an adult organism—for instance, a skin or liver cell. The next step is to try to unlock the genetic information it contains in order to reexecute the developmental program and rebuild a genetically identical organism from start. We know that such a cell contains all the genetic information of the organism, the complete genome. The scientific question is whether it can be unlocked. After all, it could be that as the developmental program unfolds and a liver cell is created, a number of irreversible transformations are introduced; for instance, genes may be removed or silenced forever. If so, the nucleus of a liver cell would not be able to function anymore as the factory where the instructions for the construction of a brand-new organism are read. The Dolly experiment showed that this is not the case, and full-blown cloning using differentiated cells from an adult organism is possible. In plants, it has been achieved since the 1950s.

Both Dolly and Cumulina (the latter, the first cloned mouse), are female and are derived from adult cells related to reproduction: Dolly, from the mammary gland cell of a ewe; and Cumulina, from cumulus cells, which nurture developing eggs inside the ovaries. One may have hypothesized that such conditions are important, if not necessary, for successful cloning. It has now been shown that both conditions are unnecessary. Wakayama and Yaganimachi (1999) used the Honolulu technique to create 274 mouse embryos from skin clipped from the tail of a male mouse. The embryos were implanted into surrogate mother mice; only three reached full term, and only one survived. Thus cloning using adult somatic cells is not restricted to female or reproductive cells. In addition, mice have been cloned from cultured stem cells.

With current technologies, artificial twinning has some advantages and disadvantages with respect to cloning obtained by nuclear transfer. Clones derived by artificial twinning are even more similar than clones obtained by nuclear transfer because, in addition to the same nucleus, they share the same initial egg. On the other hand, the genetic makeup of clones derived by artificial twinning is uncertain because it depends on a random fertilization event. This could easily change in the near future when the two techniques are combined by artificially twinning cloned embryos derived by nuclear transfer.

# Appendix C
# Stem Cells

Biological cells can be in a very large number of different developmental conditions, depending on their overall internal chemical state and their environment, which dictate which of their genes are turned on or off at any given time. Stem cells are relatively undifferentiated and can develop, under proper conditions, into different mature cell types. A cell is said to be pluripotent if it can develop into a number of, but not necessarily all, different mature cell types. It is said to be totipotent if it can develop into any cell of a given organism. It should be clear that this kind of classification is not rigid. A great deal of work is being done to determine what characterizes the developmental state of a cell at the genetic and biochemical levels, how such states can be manipulated, and whether one can manipulate a cell to move from one state to another. During development, a cell goes through a fixed, carefully orchestrated, and seemingly rigid sequence of transformations. In the traditional view of developmental biology, these transformations are not reversible. What cloning, embryonic stem cell, and other types of research are showing is that in reality cells are much more flexible than previously thought, and that with the proper manipulations they can be coerced along many possible developmental trajectories in reversible ways. In this new revolutionary view, the nucleus of a cell can be reprogrammed by the cytoplasm, which in turn is sensitive to its environment. For each kind of cell, we are progressively learning how to manipulate its environment and its components to halt its differentiation process, culture it indefinitely, and differentiate it into almost any other cell.

One of the greatest promises for the future is the use of embryonic stem cells, first obtained in humans by Thomson et al. (1998). These are cells, derived in general from an early embryo, that are capable of unlimited proliferation in vivo. This requires a medium that achieves a particular balance: not too harsh, or the cell dies; not too friendly, or the cell starts embryonic development. Embryonic stem cells can develop into virtually any cell of the body under the proper conditions (see also Solter and Gearhart 1999). (As a side issue, tumor cells are also immortal, but their DNA is often perturbed.) In time, we could learn how to induce the differentiation of embryonic stem cells and how to build artificial organs and tissues from scratch. Various other kinds of stem cells are also found in the adult (for instance, in bone marrow); and in 1999 it was shown in rats that bone marrow stem cells can differentiate into liver cells (Petersen et al. 1999), raising the possibility of one day using a patient's own marrow to repair a failing liver. Stem cells have also been used to reverse diabetes in mice (Ramiya et al. 2000).

Current research is focusing precisely on learning how to manipulate various kinds of embryonic and nonembryonic stem cells and other cell lines so that they can be both cultured indefinitely and forced to differentiate in particular directions. An achievement of Yagi et al. (1999) was to cultivate blood-producing stem cells in the laboratory. Such techniques could have a major impact on cancer treatment and bone marrow transplants. Chemotherapy and radiation treatments can kill healthy blood cells along with cancer cells. Stem cells are responsible for regrowing the blood cell system. Unfortunately, under natural conditions they are extremely rare—only 1 in 100,000 bone marrow cells is a stem cell that can produce blood. With the aid of a growth hormone discovered in 1994 (thrombopoietin) and optimized cell culture conditions, Yagi et al. generated large quantities of mouse blood stem cells that could live for up to four months outside the body. The viability of these cells was assessed by injecting them into mice that had received lethal doses of radiation; the mice recovered entirely. In the relatively near future, doctors ought to be able to remove stem cells from a patient, multiply them in the laboratory, and inject them back into the patient during chemotherapy. In a similar way, embryonic stem cells have been used in the nervous system to fight disease in rat and mice models in which fatty coating around nerve fibers (myelin) is missing.

In recent experiments, stem cells from the brains of adult mice have been shown to be capable of becoming functional blood cells. When injected into an embryo, they can even incorporate themselves into the new embryo and develop into nearly every possible tissue of the resulting chimeric mouse, including (so far) liver, spinal cord, stomach, and kidney. These experiments support the new view that cells from an adult animal can revert essentially into totipotent embryonic stem cells.

In a different kind of application, scientists were able to transplant germ cells from an infertile mouse into the testes of a mouse with a healthy testicular environment and restore fertility by producing offspring with the genetic makeup of the infertile donor male. Conceivably, similar techniques could in time be applied to humans to correct roughly 35% to 40% of infertility cases, which are due to impairments in sperm development. In addition, stem cells are easy to freeze and maintain indefinitely, and, unlike sperm, very similar freezing techniques work across different species. Furthermore, by freezing testicular stem cells, one preserves the complete genetic makeup of an individual instead of specific combinations, as is the case with sperm. Such techniques could be applied to individuals who die or who lose the ability to produce sperm before reaching puberty (for instance, as a result of cancer treatment), as well as to the preservation of endangered species.

If embryonic stem cells from a given individual were not available through cloning or some other procedure, we could imagine modifying the genomic DNA of a human embryonic stem cell to ensure that any developed tissue or organ would be compatible with the acceptor. Similarly, stem cells could conceivably be used in gene therapies to introduce normal genes into stem cells of patients affected by genetic defects. The "repaired" stem cells could then be transferred back to the patients and produce large numbers of cells carrying the corrected gene.

Because of some of the legal restrictions on stem cell research in some countries, scientists have been trying to use other, more readily available cells, such as adult stem cells or embryonic germ cells available from aborted fetal tissue (rather than embryos). These cells are derived from primordial germ cells destined to give rise during development to eggs or sperm. So far, however, results have been disappointing.

# Appendix D
# Historical and Technological Aspects of the Internet

In 1964 the Rand Corporation made public a proposal for the creation of a communication network without central authority and capable of withstanding a nuclear war in which large parts of its structure were destroyed. In 1967 the Department of Defense (DOD) proposed to connect four sites—Stanford Research Institute, the University of California at Santa Barbara, the University of California at Los Angeles, and the University of Nevada. Computers at each site could communicate with, or be programmed from, computers at any other site. This was the beginning of the ARPANET (Advanced Research Projects Agency Network), the ancestor of the Internet. The first file transfer over the ARPANET took place in 1969, using telephone wires, within weeks of man's first step on the moon on July 20. A nineteen-node demonstration of ARPANET was conducted in 1972 at the International Conference on Computer Communications. Initially, the ARPANET was primarily limited to DOD personnel, university researchers, and private contractors. In parallel to ARPANET, however, several other networks were being created, such as BITNET and USENET. Whereas ARPANET was created to survive nuclear war, these other networks were mainly used to exchange messages. Each network had its own method of communicating, or protocol; often it was incompatible with the other networks. As ARPANET evolved, a main goal became to enable communication across these disparate networks and to create an "internet"—a group of interconnected networks. In the late 1970s, work began on TCP/IP, the protocol that would be used to accomplish the first network integration. Other networks began to connect themselves to ARPANET through TCP/IP, and this was the true start of the Internet. ARPANET became the backbone of the Internet through which the other networks could talk to each other.

In particular, different technologies, such as telephone, radio, and television, were developed at different times and, initially, with an analog rather than a digital approach. The current heterogeneous state of the Internet plays a significant role in how the Internet evolves and in the commercial strategies around it. Since the 1990s there has been, for instance, an intense competition between cable television and telephone companies to connect homes to the Internet and bringing broadband (the current threshold to qualify for the term "broadband" is set at about 1 million bits per second) communication to homes. The main obstacle for broadband communication to homes is the "last mile" challenge. Unlike the backbone of the Internet, which is made of very fast optical fibers, the last mile of connection to the home was not originally designed to support broadband two-way communication.

Television cable and telephone copper wires were originally developed with different properties. Cable was meant to carry one-way broadcast analog video information, and therefore has a much larger bandwidth than telephone copper wires. Copper wires carry a narrow stream of information, fine for voice calls but too slow for the services of the future. Cable was meant to be one-directional, from broadcasting networks into homes, whereas telephone has been bidirectional for quite some time. When you turn on your television set, cable service is available immediately. In contrast, telephone service is not. It requires dialing up a specific location. Ultimately the advantages of both worlds should be available to everybody in the form of multiple broadband channels, and this is the general trend both technologically and commercially. For instance, first ISDN (integrated services digital network), and now DSL (digital subscriber lines) technologies are being used to expand greatly the amount of data that can be exchanged over telephone connections. A central telephone office is typically connected to 10,000 homes. Broadband levels are achieved by increasing the amount of optic fiber present in the "last mile" and by using better communication technology over the remaining telephone (or cable) connections. Homes are just starting to become networked (for instance, using existing telephone lines that reach almost all rooms). Soon most PCs and many appliances will come with built-in home connectivity and will be able to talk to each other. Thus we are moving toward a convergence of broadband, home and business networks or LANs (local area networks), and WANs (wide are networks) into a single gigantic network. Companies such as Terabeam are introducing new technologies to transform the last mile into the "fast" mile.

Also, the transmission rates over optical fibers will continue to increase exponentially for the foreseeable future. In March 2000, researchers at Lucent's Bell Labs managed to transmit 3.28 terabits of data per second over a distance of 300 kilometers. At these speeds, the entire Internet traffic of today could be carried on a single fiber. And this is just a beginning. Indeed, the transmission rates depend largely on the speed of the lasers used to encode the data and on the number of lasers operating at different wavelengths that can be carried by a single fiber. The number of pulses generated by a single laser currently follows Moore's law, doubling every eighteen months or so. Terabit lasers are currently under development. The number of wavelengths, or colors, a single fiber can carry simultaneously is doubling each year. At the moment, optical networks carry 80 wavelengths per fiber and 160-wavelength systems will be introduced in 2001. In the laboratory, scientists are playing with 1000-wavelength systems. Existing optical networks could be upgraded, at least in theory, to transmit $10^{15}$ bits per second. In this regard, a major technological trend is toward an all-optical network, the "worldwide web of glass and light" of George Gilder, where fast optical switching progressively replaces relatively slow electronic switching.

Another major overall trend is the move from circuits to packets. Earlier, each telephone call tied up a circuit or a pathway through the network from one phone to the other. In the Internet era, information of any kind can be broken up into packets that are reassembled at the receiving end. Many packets associated with different users can be mixed, then travel together along a single pathway. This is more efficient and, furthermore, uses a single unified IP (Internet protocol) format for all digital data. This is removing all boundaries between TVs, computers, radios, telephones, and fax machines. Different types of information may have different Internet requirements, especially in terms of timing and security, and these are essential factors to take into account in order to manage the growing complexity. Video, music, and voice transmission have stricter timing requirements than E-mail, for instance, being unable to tolerate large variable delays in the corresponding packets. Financial transactions must be secure, and so forth. Finally, digital wireless is the second driving force behind the Internet in the current communication expansion. Over 400 million mobile telephones are in use today on Earth, and this number is projected to increase by 250,000 per day and ultimately surpass the number of fixed telephones (and personal computers) in the next few years. Within ten years, we ought to see wireless phone speeds of about 2 million bits per second, allowing video transmission and cellular TV. As Gilder puts it: "The all-optical fibersphere in the center finds its complement in the wireless ethersphere on the edge of the network."

Several large software-computing projects are distributed across many programmers and computers over the Internet. We saw the SETI@home project, for the search for extraterrestrial intelligence. Several projects of this kind have appeared for cryptographic applications, or to calculate the digits of mathematical constants such as $\pi$. The PiHex project, for instance, is a distributed computing project at Simon Fraser University that uses several hundred computers to estimate $\pi$. Anyone interested in the project and with some free time on the computer can download the software and join this collective computation. Several trillion digits of $\pi$ have been computed in this way! At one site (www.distributed.net) users can volunteer computing time on their own machines and be part of the "largest computer on earth." Centrata, a recent MIT start-up, plans to capitalize on large-scale distributed computing by creating a marketplace where PC owners can sell unused computing resources. In time, we could harness the computing power of billions of cell phones and other noncomputer processors and have planetary computers.

A somewhat different example of a highly successful software project that is distributed over the Internet is the Linux operating system. Linux is a relatively simple-to-use variant of UNIX that runs, among other things, on PCs and also is frequently used in servers. Linux originally consisted of about 10,000 lines of code created in 1991 by Linus Torvalds. It now belongs to what is called the open-source software (OSS) movement, first championed by Richard Stallman at the Massachusetts Institute of Technology and the Free Software Foundation. The code is open in the sense that it is freely available to anyone, and in principle anyone can contribute to it. Naturally, any proposed modification must be approved. Since its creation, thousands of programmers around the world have been adding fixes and improvements to Linux, contributing hundred of thousands of new lines of code.

One last example is provided by Internet structures called web rings. An Internet ring is a collection of related Internet sites that organize themselves into a flexible ring where each site has a link to its predecessor and its successor, allowing visitors to navigate the corresponding information along the ring. In addition to standard information, rings have also developed around very esoteric and unusual topics.

Finally, there is some preliminary evidence that, together with economic expansion, the Internet may bring considerable energy savings, and hence environmental benefits (for instance, as a result of online shopping or telecommuting). In 1997 and 1998, the United States economy grew by about 4% while energy consumption increased very slightly. Though obviously more data points are needed to draw statistically significant conclusions, this is certainly an encouraging trend to follow.

# Appendix E
# Sizing the Internal Self Up and Other Brain Issues

In chapter 9, we defined the notion of external and internal selves. We saw how to record and measure external selves and estimate their mutual distances. One of the fundamental issues behind chapter 9 is the reproducibility of the external and internal selves, in both biological and nonbiological substrates, and the consequences for our notions of self, free will, immortality, and so forth. Although current technology suggests a number of interesting possibilities, it is very difficult to come up with definitive answers in this area. The main reason lies in the very large numbers involved, such as the size of the internal self, and our poor intuition of the physical and theoretical properties of computations in these high ranges. Here we first estimate the size of the internal self, and then list a sample of related issues that remain unclear.

Depending on the assumptions we make, the internal self requires anywhere from $10^{27}$ to $10^{40}$ bits of storage. To see this, consider the total amount of information associated with the internal electrical activity of a human brain over a lifetime. With the assumptions

- number of neurons $= 10^{12}$,
- average connectivity $= 10^{4}$,
- average information rate per axon $=$ 100 bits per second,
- number of seconds in a lifetime (70 years) $= 2.2 \times 10^{9}$ seconds,

the total number of bits per second associated with the internal activity of a human brain is on the order of $10^{18}$ bits. Over a lifetime this corresponds to a total of $2.2 \times 10^{27}$ bits.

However, there may be more to the brain and its computations than mere electrical activity. For instance, connections and synapses change as a result of activity, and so does gene expression. It is unclear at the computational level whether chemical and other changes are subsumed by action potentials. Thus, in order to define and measure the internal self, one may consider, in addition, taking a snapshot of all the proteins in a brain at regular intervals. Ultimately, at the molecular level, the computations performed by the brain are implemented by its proteins, which can interact with ligands, open and close membrane channels, modify synapses, regulate neuronal gene expression, and so forth. With the assumptions that

- volume of brain = 1.6 liters = $0.0016 \text{ m}^3$,
- volume of a protein = $100{,}000 \text{ Å}^3 = 10^{-25} \text{ m}^3$,
- total number of different proteins = 100,000 (a very rough estimate),[1]

the brain contains less than $1.6 \times 10^{22}$ protein molecules, and with 20 bits one can specify the type of protein and a few other properties. These numbers yield a protein snapshot with at most $3.2 \times 10^{23}$ bits, or a lifetime total of $7.0 \times 10^{31}$ bits, taking one snapshot per second. Although proteins can act on much faster time scales—for instance, when it comes to opening or closing ion channels—a one-second time scale seems reasonable if we address brain structural properties. This may not seem to contain the millisecond-scale information associated with action potentials, but the order of magnitude is unchanged if we add to it the previous $2.2 \times 10^{27}$ bits associated with electrical activity over a lifetime.

If we wanted to go all the way down to the level of single atoms, with the additional assumption that the volume of an atom = $1 \text{ Å}^3 = 10^{-30} \text{ m}^3$, the brain has on the order of $1.6 \times 10^{27}$ atoms. With 10 bits one can specify the type of atom (the periodic table has a little over 100 elements, for which 7 bits are sufficient, and many of which are not relevant for the brain) and a few other properties. This yields an atomic snapshot with at most $1.6 \times 10^{28}$ bits, corresponding to a lifetime total of $3.5 \times 10^{37}$ bits if the snapshot is taken every second, and $3.5 \times 10^{40}$ bits if the snapshot is taken every millisecond.

These last two large orders of magnitude ($7.0 \times 10^{31}$ and $3.5 \times 10^{40}$) correspond to 70 and 100 years of memory storage doubling, with the same, and actually even stronger, caveats stated in chapter 9. With an internal self sized at $10^{31}$ and a human population of 10 billion, one would need a combined storage capacity of $10^{41}$ bits to be fair to everybody. For comparison, the number of particles on Earth is in the range of $10^{50}$ and the number of particles in the universe is believed to be in the $10^{90}$ range.

In terms of reproducibility, how similar can the external selves of two distinct human beings be? Even if perfect identity is not possible, can we get arbitrarily close to it, or is there a minimal radius that cannot be surpassed? Such a radius could depend on the individual, and could have many sources, such as

- High sensitivity of neuronal circuits, including the possibility of chaotic behavior whereby even small variations in initial conditions tend to be amplified over time
- The existence of a spectrum of random effects (for instance, at the atomic and molecular levels) during development and so on (this is distinct from the previous point because chaos can exist in deterministic systems)
- The "curse of dimensionality," essentially the fact that it is hard to sample high-dimensional spaces. Producing 1000 clones of the same individual and raising them in similar conditions in an attempt to build a perfect copy may provide only a very sparse sample of external selves with similar genomes. Identical twins can have different, although often correlated, fingerprint patterns. Similarly, details of brain anatomy of two clones must include plenty of differences, some arising early in development. Fine-grained details of brain anatomy cannot be specified in the relatively small amount of DNA we carry.

Thus it is likely that such a radius exists at least for all practical reasons, if not for theoretical ones. In fact, we have seen that in practice it already exists for PCs, and is likely to increase in size with the complexity of the computing device. But the point is that such a radius in principle is considerably smaller than what we have been tricked to think during our evolutionary history. For instance, one ought to be able to produce individuals that pass increasingly sophisticated forms of the personalized form of the Turing test; that is, individuals that are indistinguishable in a conversation. Would it be easy to distinguish Oliver Sacks's twins in a conversation? A fortiori such a radius ought to exist for internal selves that are several orders of magnitude more complex, and problematic to measure and compare.

In a similar vein, how close can the input-output function of a nonbiological device be to the input-output function of a biological brain? We have seen that to some extent one could envision a machine-learning approach to the problems using, for instance, artificial neural networks. That is, we could envision using the input-output function of a brain over an interval of time $[0,t_1]$ to train a gigantic artificial neural network at some time $t_2$ greater than $t_1$. Several additional issues that are unclear include the following:

- How long would training take, and how close could $t_2$ be to $t_1$?
- More generally, what are the complexities of this learning process and its statistical boundaries? Learning from examples is inherently a probabilistic process: it does not always succeed, and even when it does, it always includes some generalization error. What is the time horizon beyond $t_1$ over which generalization can be assessed, and with which input?
- The brain-equivalent artificial neural network is not a feedforward network but a recurrent one replete with feedback cycles. Learning in recurrent networks is notoriously slower and more difficult, and existing algorithms would scale poorly to the task.
- The issue of analog versus discrete time and computations: Should the artificial neural network reproduce the precise timing of spikes, or just average spiking rates? How much precision is required on the inputs, outputs, and synaptic weights? Should the artificial network have a discrete clock and compute at discrete intervals? What is the relation between the speed of computation in the artificial and biological networks?
- From the point of view of parameters, the artificial brainlike networks we discussed in chapter 9 contain on the order of $10^{22}$ synaptic weight parameters. A seventy-year lifetime of external self captured every millisecond would give only the equivalent of $2.2 \times 10^{12}$ input-output training pairs. Thus the number of constraints is very small with respect to the number of free parameters, and is likely to result in a high degree of indeterminacy. The external self alone does not contain enough information to uniquely determine the artificial internal self. This situation could be mitigated in part by the existence of structural constraints, such as modularity, but it is impossible to evaluate the trade-offs between the two with current knowledge.
- Viewing the brain as an input-output device is highly deceiving for several reasons. First of all, the device itself is not stationary but varies in time. Neuronal circuits change their structure, and many learning and

other computational processes take place at the chemical level of synapses or DNA regulation, and are not described by the electrical activity or the input-output function alone. Second, the output influences the input—we move our eyes to change what we see.

And this is without touching issues of whether the brain equivalent has any kind of self-awareness, any notion of time, or the ability to detect whether its electric power has been turned on or off. The large numbers involved suggest that it is extremely unlikely that in the foreseeable future one could build a computer capable of predicting the future evolution of a brain, even over a short time scale. Thus, for all practical purposes our behavior will remain inherently unpredictable for the foreseeable future, which is consistent with, but not necessarily identical to, our sense of free will. In this sense, free will is computational. But is it purely computational? Whether this is a temporary situation or corresponds to a fundamental limitation of our universe remains to be seen. In a finite universe, there is a finite amount of computing power available at any given time, and therefore a computational system beyond a certain level of complexity cannot be emulated in real time by any other computational system. How close human brains are to this threshold remains to be seen.[2]

# Notes

## Chapter 2

1. Crichton added: "*Under the circumstances, for scientists to fret over their image seems absurd.*" But of course science is also very much a human enterprise where politics, history, society, and human characters play important and complex roles. This is seen in philosopher Paul Feyerabend's words: "*Success in science depends not only on rational argument but also on a mixture of subterfuge, rhetoric, and propaganda,*" a sentiment in part echoed by Sydney Brenner, a well-known molecular biologist: "*It is only through the use of subterfuge such as applying for money for work already done that innovative research can be freely pursued.*"

2. French mathematician Pierre de Fermat's last theorem is a simple statement about the equation $a^n + b^n = c^n$. It states that when n is greater than 2, the equation has no nontrivial integer solutions. When $n = 2$, there are many solutions, such as $a = 3$, $b = 4$, $c = 5$. Trivial solutions are obtained when a, b, and c are all equal to 0 or 1. During the seventeenth century, Fermat wrote the "theorem" in the margin of a book. He stated: "I have discovered a truly wonderful proof that this margin is too small to contain." For over three centuries mathematicians tried to prove or disprove the statement. During the 1900 International Congress of Mathematicians in Paris, German mathematician David Hilbert gave an influential address in which he listed twenty-three major open mathematical problems to be studied in the coming century. Fermat's last theorem was one of them. It was finally proved by Andrew Wiles in 1994. His original proof, presented for the first time at the Newton Institute in Cambridge, England, in 1993, contained an error which Wiles fixed later that year.

3. Note that some problems are *not* solvable. A classic example is the Halting Problem: deciding whether, given certain input, a computer program will stop after any time. Intuitively, this is not solvable because if we let the computer run for any finite length of time, say ten days, after which it has not stopped, we cannot draw any reliable conclusion. It could very well stop on the eleventh day. In the

1930s mathematicians such as Kurt Gödel and Alan Turing showed that the Halting Problem is undecidable, or not solvable. The P = NP problem goes back to a paper presented in 1971 by the mathematician Stephen Cook, "The Complexity of Theorem Proving Procedures."

4. With a fixed starting city, the number of tours is given by $(N - 1)! = 1 \times 2 \times 3 \times 4 \times \ldots \times (N - 1)$. Thus there are 2 tours for 3 cities, 6 tours for 4 cities, 24 tours for 5 cities, 120 tours for 6 cities, and so forth.

5. The notion of "efficient" should be formalized. "Efficient" means that the solution of a problem with input size N should take an amount of time that is upper-bounded by a polynomial function of N, which is equivalent to saying N multiplied by itself a certain fixed number of times. It can be shown that a large class of combinatorial decision problems are equivalent to each other, and to the Traveling Salesman Problem, in the sense that any problem in the class can be encoded into any other problem in the class, using a polynomial time encoding. Because a polynomial of polynomial variables is still a polynomial, it is clear that if one problem in the class has a polynomial time solution, then all the other problems in the class also have a polynomial time solution. To be precise, as stated (finding the shortest tour), the Traveling Salesman Problem is broader than a simple decision problem. But it is easy to provide a decision version of the problem by asking whether there is a tour of length less than K, for some K.

6. One area of physics in which fundamental principles are still involved is the search for a grand unified theory (GUT) that would unify the four known forces in the universe, from large-scale gravitation, to electromagnetic, to short-range weak, to strong nuclear forces. As fascinating as such a theory might be, it is unlikely that it would have any direct impact on the world as we observe and manipulate it. Most important for this book, it would have little effect on biology.

7. The first solid clue in this direction came in 1828 with the synthesis of urea by Friedrich Wohler. This was the first synthesis of an organic compound, known to be produced only by living organisms, starting from scratch using only inorganic elements. This discovery dealt one of the major early blows against vitalism. Today we know that entire organisms are built from a few wholly inorganic starting materials, although the assembly process may be exceedingly complex.

8. In fact, the mathematical formalization of this inference process, which consists in generating hypotheses, and assessing and refining them through experiments, is even more recent. It can be traced to the work of astronomers such as Pierre-Simon de Laplace, but was formalized by scientists only during the twentieth century, in what is called Bayesian statistics.

9. In fact, decentering can also be followed throughout the history of mathematics, with increasing levels of abstraction. For instance, in the movement that took us from a handful of numbers to integers, to rational numbers, to real numbers, to complex numbers, from small values to asymptotic expansions, from one-dimensional geometry to n-dimensional geometry, from Euclidean to non-Euclidean geometries, and so on.

10. For instance, the total number of particles in the visible universe is currently estimated at $10^{90}$. Most of these are light particles, or photons, that make up the cosmic background radiation together with a comparable number of neutrinos, subnuclear particles without electric charge.

11. More formally, this approach is used in a branch of physics called order-of-magnitude physics and aims at obtaining quick estimates for just about any imaginable quantity by using common sense, the way physical laws scale with respect to each other, and a few other similar tricks. In spite of its simplicity, this approach can lead to remarkable and sometimes surprising results. The most important thing in any order-of-magnitude calculation is to keep a clear picture of the assumptions on which it relies and the final goals. For instance, $10^{10}$ may seem a large quantity, but it is completely irrelevant compared with $10^{20}$. As a result, in order-of-magnitude algebra, $10^{10} + 10^{20} = 10^{20}$.

12. This is a slight underestimate, because the volume of the typical human brain is more like 1.6 liters.

13. It is an upper bound because we neglected the volume occupied by axons and dendritic trees—the elements interconnecting neurons—the space between neurons, and the presence of other supporting but noncomputing cells such as glial cells, and we used a crude lower bound on the volume of a single neuron. If we estimate the typical radius of a neuron to be at least 3 microns, then we immediately get a revised upper bound of $10^{13}$. Taking into account the other factors gives a range closer to $10^{12}$. Alternatively, we can divide the 1.6-liter brain volume by a more realistic estimate of 1000 cubic microns for the volume of the average neuron, to get again the estimate of $1.6 \times 10^{12}$ neurons. The main arguments made in the book, however, are robust and do not require more precision. Whether the total number of neurons is closer to $10^{12}$ or $10^{14}$ is not essential for the points to be made. Similarly, the total number of humans who ever lived is most likely less than 100 billion, and probably closer to 10 billion. But whether it is 10, 50, or 200 billion does not fundamentally change our conclusions.

## Chapter 3

1. Many other molecules are necessary for life: from oxygen and metals, to water, to more complex organic compounds such as most vitamins, lipids, and carbohydrates. These molecules interact with DNA/RNA and proteins. Without them, DNA/RNA and proteins could not function. But it is fair to say that DNA/RNA and proteins are the key "intelligent" players.

2. The four bases are adenine, cytosine, guanine, and thymine. In the case of RNA sequences, the letter T is replaced by the letter U (uracil).

3. There are some exceptions, as is always the case in biology. Some genes code for RNA elements.

4. One of the main theories concerning the origin of life on Earth is that close to the very beginning, life consisted of RNA-based, self-replicating molecular

assemblies. This is because, unlike DNA, RNA can both catalyze chemical reactions and store genetic information. Several scientists have even suggested that RNA-based, independently replicating descendant organisms may still exist, hidden somewhere on our planet, perhaps at the bottom of an underground Russian lake. However, nobody has found such organisms so far, nor have we been able to produce any kind of self-replicating molecular assemblies in the laboratory. What we have known for quite some time is that organic molecules and even amino acids arise spontaneously in the laboratory equivalent of a primordial soup energized by a few electrical sparks. Recently, DNA enzymes—short DNA sequences with some enzymatic activity—have also been found. But determining the precise chemical origin of life remains a difficult problem.

5. The twenty amino acids, and their letter codes with three letters or one letter

| | |
|---|---|
| Alanine (Ala or A) | Leucine (Leu or L) |
| Arginine (Arg or R) | Lysine (Lys or K) |
| Asparagine (Asn or N) | Methionine (Met or M) |
| Aspartic Acid (Asp or D) | Phenylalanine (Phe or F) |
| Cysteine (Cys or C) | Proline (Pro or P) |
| Glutamine (Gln or Q) | Serine (Ser or S) |
| Glutamic Acid (Glu or E) | Threonine (Thr or T) |
| Glycine (Gly or G) | Tryptophan (Trp or W) |
| Histidine (His or H) | Tyrosine (Tyr or Y) |
| Isoleucine (Ile or I) | Valine (Val or V) |

Hundreds of other amino acids exist that are not used in the makeup of living organisms. Scientists are working on creating new proteins, based on these amino acids, with new chemical properties. This can be attempted by engineering new forms of DNA with five letters instead of four, or with a different genetic code. Ultimately, by reengineering DNA, RNA, and proteins, it may be possible to create novel forms of life.

6. The genetic code associates any triplet of DNA bases with an amino acid or a stop signal

| | | | |
|---|---|---|---|
| TTT = Phe | TCT = Ser | TAT = Tyr | TGT = Cys |
| TTC = Phe | TCC = Ser | TAC = Tyr | TGC = Cys |
| TTA = Leu | TCA = Ser | TAA = Stop | TGA = Stop |
| TTG = Leu | TCG = Ser | TAG = Stop | TGG = Trp |
| CTT = Leu | CCT = Pro | CAT = His | CGT = Arg |
| CTC = Leu | CCC = Pro | CAC = His | CGC = Arg |
| CTA = Leu | CCA = Pro | CAA = Gln | CGA = Arg |
| CTG = Leu | CCG = Pro | CAG = Gln | CGG = Arg |

| | | | |
|---|---|---|---|
| ATT = Ile | ACT = Thr | AAT = Asn | AGT = Ser |
| ATC = Ile | ACC = Thr | AAC = Asn | AGC = Ser |
| ATA = Ile | ACA = Thr | AAA = Lys | AGA = Arg |
| ATG = Met | ACG = Thr | AAG = Lys | AGG = Arg |
| GTT = Val | GCT = Ala | GAT = Asp | GGT = Gly |
| GTC = Val | GCC = Ala | GAC = Asp | GGC = Gly |
| GTA = Val | GCA = Ala | GAA = Glu | GGA = Gly |
| GTG = Val | GCG = Ala | GAG = Glu | GGG = Gly |

7. Chromatin fibers have a lace-of-pearl structure. The pearls are nucleosome particles, each of which consists of 145 bases of DNA wrapped around proteins called histones.

8. The HIV genome is actually made of RNA, which uses uracil (U) rather than thymine (T).

9. Viruses, however, are not independent, self-replicating entities—they need a host. The smallest known independent self-replicating cell is *Mycoplasma genitalium*. Its genomic sequence contains 517 genes. Recent evidence seems to indicate that only 265 to 350 protein-coding genes are essential under permissive laboratory growth condition. One line of research is trying to determine the smallest complements of genes, or the shortest "software" programs, needed by an independent self-replicating cell. The next grand challenge in this direction would be to assemble such a cell more or less from scratch, not only the genomic DNA but also all the proteins and other components, to create life from its chemical components.

10. All the base pair numbers given are for the haploid genome corresponding to one chromosome in each homologous pair. As the rush to the finish line of the Human Genome Project is taking place, estimates of the total number of genes vary wildly from 30,000 to 140,000. Chromosome 21, the second human chromosome to be sequenced, contains less than 300 genes consistently, with a total estimate in the range of 40,000 genes, many fewer than previously thought.

11. Each chromosome contains roughly 4000 genes and 125 million base pairs. This is a blowup by a factor of 10 million ($10^7$).

12. Needless to say, a great deal of variability is found at each scale. Smaller players, such as electrons, are also very important. Photosynthesis, respiration, drug metabolism, DNA repair, and many other biological and chemical processes depend on controlled electron flow through proteins. These reactions are remarkably rapid and specific in spite of the large molecular distances (often greater than 10 Å) between electron donors and acceptors.

13. For reference and example, the size of sodium ions ($Na^+$) in crystal form is 1.33 Å, and of potassium ions ($K^+$) is 0.95 Å.

14. Before meiosis, in the cell that produces eggs or sperm, each pair of chromosomes is pressed together and may swap portions of genetic material. This form of recombination is called crossing over. It is very frequent, occurring at least two to three times per chromosome per meiosis. Thus, if you look, for instance, at the pair of chromosomes 5 you have in each nonsexual cell of your body, the chromosome contributed by your father contains a combination of genetic material originated from your paternal grandfather and grandmother. This combination was created by crossover during the meiosis of the single spermatozoid that later fertilized the egg you resulted from. Crossover, mutations, and independent assortment are stochastic events responsible, under natural conditions, for the diversity, uniqueness, and evolution of genomes. These are not the only genetic events, though; there are other phenomena, such as lateral transfers of genetic material from one species to another (for instance, by viruses).

15. In fact, to preserve the dichotomy, genes and their products must usually be patented in a form that is not found in nature—purified, inserted in a different organism, and so forth. As we shall see, in the long run such a distinction is problematic.

16. In 1998, each machine was able to process over 750 samples per day with less than an hour of human labor. Assuming a typical length of about 400 DNA bases per sample, and coverage of about ten times for each position in the human genome, this works out to about 75 million samples, or roughly 340 days. After helping to complete the fruit fly's genome, Celera began sequencing the human genome on September 9, 1999. On October 20, 1999, it announced that it had sequenced and delivered to its subscribers approximately 1.2 billion bases of human DNA, roughly one-third of the entire genome. Celera also announced that its scientists had been able to identify a large number of new gene fragments. According to Dr. J. Craig Venter, president and chief scientific officer of Celera: "Included in this set are several thousand receptors, ion channels, and secreted proteins that we believe will be of significant interest to pharmaceutical companies." Celera filed provisional patent applications on 6500 of these discoveries. On April 6, 2000, it announced the completion of the sequencing phase of the genome from one human being and started sequencing the mouse genome. The sequencing phase was followed by an assembly phase in which mathematical algorithms and a supercomputer are used to assemble the fragments over a period of a few weeks, as in a gigantic jigsaw puzzle. Celera also announced plans to study proteins on a massive scale, using another fleet of machines, such as mass spectrometers, and its $100 million supercomputer, which, at the time of this writing, already held over 70 terabytes of genomic data, several times the contents of the entire Library of Congress. Several other companies, such as Incyte and Human Genome Sciences, are involved in various aspects of large-scale sequencing and bioinformatics. The race and fights between the private and public human genome sequencing efforts were ongoing at the time of this writing, making it impossible to compare their status objectively. But, if nothing else, the original announcement by Celera accelerated the academic human genome sequencing efforts aimed primarily at producing *public* data by several years.

17. Sequence analysis is now one of the most powerful tools through which evolution can be studied, from the origin of life on Earth to the most recent human history and even to contemporary events. It is being used, for instance, to investigate questions ranging from the origin of the X and Y chromosomes, to whether all modern humans share a single common ancestor, to the movement of human populations around the globe over the last 10,000 years, to the contemporary evolution patterns of the AIDS and influenza viruses. Another well-known application of sequence analysis is in forensics.

18. The expression begins with the transcription of the DNA template into a corresponding single-stranded RNA chain called messenger RNA (mRNA). In higher organisms, the genetic DNA information remains in the nucleus of the cell. The mRNA, however, travels outside the nucleus to the cytoplasm, where it interacts with small protein factories called ribosomes, in which mRNA is translated into protein. The structure of ribosomes was resolved in 1999.

19. For instance, signal peptides or inteins. Signal peptides are short extensions of about fifteen to thirty amino acids attached to secreted proteins such as antibodies (immunoglobulins). They direct newly synthesized proteins and are cleaved off before secretion of the corresponding mature protein. An intein is a protein sequence embedded within a precursor protein sequence and spliced out during a maturation process, termed protein splicing. Protein splicing involves precise excision of the intein sequence and joining of the flanking sequences (called exteins).

20. Thus the regulatory circuit can be simplistically viewed as a sparse circuit with 40,000 nodes in which most nodes interact with a relatively small number of other nodes. In fact, the notion of networks that are both sparse and hierarchically organized seems to be a common theme of complex systems in biology and other areas. In the brain, for instance, a given neuron might be connected to as many as 100,000 other neurons—still a very small number with respect to the total number of neurons. Elementary operations in such a system are implemented by the interaction or communication of a node with its neighbors.

21. These are respectively called transcriptome and proteome. The differences come from the fact that there are many regulatory steps between the expression of a gene and the production of a protein. For a particular gene at a particular time, a high level of transcription does not necessarily correlate with a high level of expression for the corresponding protein.

22. The same combinatorial principles can be used to determine sequences, SNPs, protein levels, or how a protein interacts with a large number of other compounds in high-throughput drug screening. Visualization of both DNA and protein levels—transcriptome and proteome—can provide further insights into regulatory mechanisms.

23. This is also the company that successfully cloned the first pigs by nuclear transfer in March 2000.

24. In the case of the firefly, it is the gene coding for the enzyme luciferase. This enzyme breaks down the chemical luciferin, which can be added to the water

given to the plant. When luciferin breaks down, light is released in the process. This results in plants that can glow in the dark. The GFP (green fluorescent protein) gene was isolated in 1991 and is routinely used as a genetic marker. The gene codes for a jellyfish protein that induces green fluorescence. A transgenic fluorescent mouse was created in 1997 by a Japanese team (Okabe et al. 1997).

In 1999, a team in Edinburgh, Scotland, was able to engineer a potato with the jellyfish protein that glows when it needs watering. Such potatoes do not have to enter the human food chain, but could be used as sentinels in food crops. Eight plants per hectare would allow a farmer to monitor an entire field. The glow is not visible to the human eye but is easily seen with a small detector. It has been estimated that the system could save up to $400.00 per hectare in terms of reduced levels of water and fertilizers. In time, similar techniques could be developed to reveal the status of various other nutrients and further optimize agricultural practices.

25. Gene therapy has already been used successfully to treat hemophilia B in dogs and humans (Herzog et al. 1999; Kay et al. 2000). It is a rare blood disorder that affects about 5000 Americans, caused by insufficient levels of a clotting protein called Factor IX. A healthy copy of the corresponding gene was delivered, using the virus AAV as a vector. Another recent major breakthrough is the treatment of severe combined immunodeficiency (Cavazzana-Calvo et al. 2000), using stem cells.

As far as more conventional therapies are concerned, the total number of drugs available today is relatively small. Present-day drugs address only on the order of 500 molecular targets. In the relatively near future, there is room for a number of targets, and hence of drugs, at least one order of magnitude larger.

26. Another surprising bacterium is *Deinococcus radiodurans*. It can survive exposure to more than 1000 times the amount of radiation that would kill a human. It can even strive when exposed to radiation continuously. It was discovered by food scientists in 1956, in cans of meat that had spoiled in spite of supposedly sterilizing irradiation. After being hit by strong radiation that shatters its DNA into hundreds of fragments, *D. radiodurans* is capable of stitching its genome back together within twenty-four hours. By comparison, a common bacterium such as *E. coli* in general cannot survive more than two or three complete breaks in its DNA. The *D. radiodurans* genome was recently sequenced and contains about 3.1 million base pairs. Each bacterium contains from four to ten copies of its genome, increasing the odds that a damaged gene will have an undamaged counterpart. Scientists are trying to bioengineer new strains of the bacterium that could be used to biodegrade polluted radioactive sites. To date, the Department of Energy has disclosed about 3000 sites contaminated by weapons production and development of nuclear reactors. Many of these sites contain a mix of organic pollutants such as trichloroethylene, radioactive elements such as uranium and plutonium, and several heavy metals.

27. In particular, mitochondria are one type of organelle—the powerhouse of the cell, where energy is fabricated—that is always inherited from the mother. Mito-

chondria also contain some DNA, the function of which is not totally understood. This DNA is believed to be more stable than the genomic DNA, and therefore has often been used in genetic studies of humans, although this aspect is still somewhat controversial (see Strauss 1999a and the related articles in the same issue of *Science* starting on p. 1475).

There are also other forms of inheritance. For instance, in what is called epigenetic inheritance, heritable changes in gene expression occur without a change in the DNA sequence. Indeed, we may inherit the control state of particular genes—for instance, in the form of silenced genes at the chemical level—in addition to their DNA sequence. This is because chemical groups can be added to or removed from DNA to limit or increase its accessibility to proteins, such as transcription factors. Thus two DNA sequences or two genomes that are identical at the primary sequence level could lead to different patterns of gene expression. The precise extent of epigenetic inheritance in humans is not well known, but there is no reason to believe it could not be understood and even manipulated in the not too distant future. Finally, there is the nongenomic transmission of behavior and responses (see, e.g., Francis et al. 1999).

28. The analogy between genomes and computer programs can be pushed even farther. For instance, one widely used technique to protect proprietary programs is to introduce a "time bomb," a mechanism that prevents a program from being copied or used after a certain date or event. One equivalent gene-protection technology is the sterile seed technology, whereby seeds give rise to plants that are sterile. Such seeds can be used only once, forcing users to purchase new sets for each year's crop. Over a dozen companies and public institutes have at least thirty-one patents with claims involving seed sterilization. Under public and private pressure, in 1999 two biotechnology firms, Monsanto (whose sterile seed technology was dubbed Terminator) and United Kingdom-based Astra Zeneca, officially announced that they would not commercialize sterile seed technology. Related gene-protection techniques, in which only value-added traits are deactivated from one generation to the next, are also being investigated.

## Chapter 4

1. In fact ectopic pregnancies, pregnancies outside the womb (but still inside the mother), accidentally happen in humans. They are very rare and typically result in the mother's death unless they are terminated. In a few cases, however, ectopic babies were delivered by cesarean section. In September 1999, in England, a woman delivered triplets, two who whom had developed in her womb, and the third in a fallopian tube.

2. It is conceivable, for instance, that one could find ways to accelerate development and reduce the traditional nine-month period.

3. In a 1999 Internet hoax, Sony was depicted as having developed the first prenatal television for children in the womb, the Utertron 9000, with programs specially tailored to each trimester of pregnancy.

4. In principle, initial experiments could be conducted on animals to study under which conditions a womb could be positioned surgically outside the body of the mother, and later maintained alive by artificial means only.

5. In a sense, this is happening now as a result of artificial and premature deliveries. A baby with a large head that would have endangered its survival in a normal delivery can be delivered more safely with modern technology. Later, it can pass the corresponding genes to offspring who will also have a lower chance of being delivered vaginally.

6. Implanting a fertilized egg into a surrogate mother is another technique, already available today, that removes the need for pregnancy by displacing it onto another woman. The theme of surrogate motherhood and surrogate mother slaves is exploited in Margaret Atwood's novel *The Handmaid's Tale*, in which a fertile woman is forced to serve as a baby machine for a commander whose wife has become sterile as a result of an ecological disaster.

7. This cannot be done using the digestive system, because digestive acids would destroy the genetic information, or using the respiratory system. These are the main systems available for any material exchanges with the external world.

8. Similar adaptations to avoid desiccation are found at the level of embryonic development. Fish embryos develop directly in water. For birds and reptiles, a hard-shelled egg protects embryonic development. In mammals, embryonic development must take place in a cavity within the body of the mother. In the case of fish embryos, metabolic waste can be released directly into the environment. In birds and reptiles, the egg preserves the water content, but waste and nutrients must be compartmentalized and recycled. Mammals have a special exchange system between the mother and the embryo (the placenta), because the mother processes all nutrients and waste. Thus there is a gradual evolutionary adaptation to dry land from fish, to amphibians, to reptiles and birds, to mammals.

9. See Schulz et al. (1999) for a detailed anatomical study in which willing couples were asked to perform in a Magnetic Resonance Imaging (MRI) machine, for "it has been extremely difficult to investigate anatomical changes during the act of coitus." One of the main results: "During intercourse in the 'missionary position' the penis is neither straight nor 'S' shaped, as had been previously thought, but is, in fact, the shape of a boomerang."

## Chapter 5

1. This discomfort has not gone unnoticed by the advertising industry. During a relatively short period after the Dolly announcement, many magazine, newspaper, and TV advertisements contained explicit references to cloning in one form or another, in order to elicit a feeling of discomfort in the audience and to promote the "individuality" of certain products.

2. In addition to use of cloning in advertisements, in June 1999 the first issue of a new quarterly scientific journal dedicated entirely to the topic of cloning was published: *Cloning*, edited by Ian Wilmut.

3. The story of a forbidden romantic situation between a brother and a sister that ends up being happily resolved by the finding that the two in reality are not genetically related (because of the philandering of their respective parents) has been used in the literature—for instance, in the work of the seventeenth-century playwright Molière.

4. The theme of human selection by beauty has been explored in books and films, including Boris Vian's *And We Shall Kill All the Ugly Ones* (*Et on tuera tous les affreux*) and the movie *Moonraker*, starring Roger Moore as James Bond.

5. In many ways, this is the theme behind the 1996 film *Multiplicity*, in which a man overburdened with too many tasks decides to clone himself to relieve some of the pressures of everyday life. But his clones clone themselves, which does not simplify the life of the man's wife. Human cloning is also a major theme in the 2000 movie *The 6th Day*, starring Arnold Schwarzenegger.

6. A theme in Michael Apted's film *Nell*, starring Jodie Foster.

7. Oliver Sacks recounts how he found them intensely immersed in a sort of holy game: "they were seated in a corner together, with a mysterious, secret smile on their faces, a smile I had never seen before, enjoying the strange pleasure and peace they now seemed to have. . . . They seemed to be locked in a singular, purely numerical converse. John would say a number—a six-figure number. Michael would catch the number, nod, smile and seem to savor it. Then he, in turn, would say another six-figure number, and now it was John who received, and appreciated it richly. They looked, at first, like two connoisseurs wine-tasting, sharing rare tastes, rare appreciations" (p. 201). He reports, after checking, that the numbers they were throwing at each other were all prime numbers. In fact, on a subsequent occasion, he joined the game with the help of a table, and even helped extend the game toward longer primes, to their great pleasure. By the time he left, they seemed to be dealing with twenty-figure prime numbers: "numbers are not just awesome to them, they are friends too—perhaps the only friends they have known in their isolated, autistic lives" (p. 208). In a different incident that formed the basis of a scene in the film *Rain Man*, when a box of matches was spilled on the floor, the twins simultaneously cried "111," the correct number of spilled matches.

We have to take Oliver Sacks at his word. It is clear, however, that valuable information is missing and the twins could have been studied further. For instance, in the incident with the matches, one would love to know how repeatable it is, what is the largest number of matches they can handle, what is the degree of precision, and the length of time it takes for them to produce an answer. The duration, for instance, could provide some crude indication of the underlying processing: whether it is in part serial or entirely parallel, and so forth. By comparison, the largest number of objects normal individuals can reliably count during a very brief presentation is only on the order of five to seven. This seems to be achieved in parallel, and any enumeration beyond this level seems to engage some kind of serial processing. The same is true for the recognition of large primes or for the calendrical algorithm. How does the response time increase with the length of the prime or the remoteness of the date? It might have been difficult to test the twins in any systematic way, and certain computer or psychophysics ideas may have been less obvious in the 1960s. But regardless of these technical points, one senses that something exceedingly mysterious is at work in the twins. They represent something both very fundamental and very rare under natural conditions that we do not understand.

None of the explanations suggested for the twins' strange powers is very convincing. For the calendar problem, one author speaks of a simple solution by which one divides the total number of days between "now" and "then" by 7. But how does one see the total number of days between now and February 14 of the year 44444? And how can the twins' brains implement an unconscious division by 7 and visualize the result? Sacks quoted a study saying, "Something mysterious, though commonplace, is operating here—the mysterious human ability to form unconscious algorithms on the basis of examples" (p. 198). In my judgment, it is very doubtful that what we observe here results predominantly from simple learning through examples alone. Although we do not know the details of the twins' history, it is hard to believe that they were fed with many prime numbers or the day of the week of many distant dates during their childhood. It also seems unlikely that a "normal" person, even with massive amounts of training, would readily learn how to recognize ten-digit primes or count, almost instantaneously, a large number of identical objects in a visual scene. Rather, it seems more likely that the twins had a genetic makeup resulting in very peculiar brains with a strong predisposition, if not inevitability, for certain types of operations. Perhaps the ability to store inordinate amounts of information in visual memory, perhaps the ability to conduct deep-nested recursions—who knows?

8. Under natural conditions, the rate for identical, or monozygotic, twins is about 4 per 1,000 births and is relatively constant over time and across human populations. The rate for nonidentical or dizygotic twins is about 2 per 1,000 births. This rate, however, has been growing steadily in recent years as a result of the use of fertility drugs—ovulation-inducing hormones—resulting in more eggs being produced and fertilized, possibly in vitro. In Japan, for instance, the dizygotic twin rate had remained nearly constant from 1975 (1.86) to 1986 (2.27). By 1994 it had increased to 3.89 per 1,000 births.

9. This crude estimate assumes independence, and might be even lower if one takes into account some of the known weaknesses associated with identical twins produced under natural conditions.

10. Of course, a large population of genetically identical individuals may be vulnerable to a new type of virus or bacterium, but this is not at all the point I wish to make. What matters in this thought experiment is the ability to continuously generate new samples from the population. And one could imagine doing so under sterile conditions or with a level of medical sophistication so that viral and bacterial infections would not be a problem anymore. And a genetically uniform population, with a uniform response to drugs and vaccines, could potentially make things easier from a medical standpoint.

11. If the DNA at the end of the chromosomes of adult somatic cells shrinks a little after each cell division, it is clear that a clone created from an adult cell has less telomeric DNA at his or her starting point. Furthermore, if such a mechanism is indeed involved in the aging process, it could be that a clone derived from an

adult cell ages faster. Indeed, shorter telomeric DNA was recently reported in Dolly. Opposite results, however, have been reported in calves and mice. In calves, for instance, cells that were allowed to divide and age as much as possible were used to clone six calves. The cloned calves had telomeres longer than normal cattle the same age. This and other experiments seem to show that under the proper conditions, cloning can turn back the clock and perhaps even increase life span. Cows can live to about twenty, however, so we will not have a sure answer for at least two more decades. We are just beginning to be able to test such hypotheses experimentally. The difference between Dolly and the cloned calves could have several origins, including random variations, differences in nuclear transfer methods, differences between cell types, and differences between species.

There are several other theories of aging. Aging may result from the accumulation of mutations and damage to mitochondrial DNA—the DNA contained in the organelles, where the energy of the cell is produced—leading to a progressive reduction of the energy available to the cell to sustain its processes. Others have hypothesized that aging may be associated with the progressive deterioration of the regulatory circuitry associated with a small number of key genes. Most likely, aging is a complex phenomenon with multiple coexisting facets.

Another issue is how much evolutionary pressure operates on aging. Though at first sight aging and death might seem beneficial to the species, it is also possible that aging may be nonadaptive. In the pleiotropic theory, aging results from a decrease in the force of natural selection with age. In this case, aging could result, for instance, from detrimental effects late in life of genes that act beneficially in early stages of life.

12. A different set of issues is whether this is desirable or economically feasible, and how society ought to evolve to accommodate such changes. Would you like to be married to the same person, or hold the same job, for 150 years?

13. Xenotransplantation raises the concern that humans might become infected with animal viruses. Such concern is heightened by the fact that the AIDS retrovirus is believed to have spread from monkeys to humans. A recent study of 160 patients who received living pig tissues, however, did not find any evidence for transmission of porcine endogenous retroviruses. Surprisingly, the study found that pig cells remained alive in humans for much longer than previously thought, up to 8.5 years. Although these results are encouraging, further testing is required.

14. There is a worldwide shortage of organs. At the beginning of 2000, over 100,000 people in the United States and Europe were waiting for livers, kidneys, and hearts. After the birth of the first five cloned piglets, Ron James, managing director of PPL Therapeutics, said, "An end to the chronic organ shortage is now in sight."

On a different note, showing one more complex interaction between biology and the Internet, the commerce in human eggs has expanded over the Internet. In September 1999, a message for the auction of human body parts briefly appeared on the Ebay web site, then was shut down by the company, which was forced to restate its policy forbidding the sale of body parts.

15. The fascination with the degree of identity between twins was reflected in a recent BBC series, *The Secret Life of Twins*, and an exhibit of photographs of twins at

the Wellcome Trusts's Two10 Gallery in London. The photographs are by Chicago cosmetic surgeon David Teplica, who is quoted as saying, "It's amazing to . . . see things like two 12-year-old [identical twin] girls whose acne has erupted in exactly the same spot on the nose."

16. Depending on the cloning technology used, two clones may or may not share the same mitochondrial DNA. In the case of embryonic splitting, or artificial twinning, the mitochondrial DNA is the same. Dolly the sheep, on the other hand, is a genetic chimera. The genomic DNA in the nucleus of its cells is equal to the nuclear DNA (donor) of the animal originally cloned, whereas its mitochondrial cytoplasmic DNA is derived from the oocyte (egg host) used during the cloning operation.

17. The theme of cloning and human farming has been exploited in the literature—for example, in Aldous Huxley's novel *Brave New World* and Boris Vian's *And We Shall Kill All the Ugly Ones* as well as in countless fiction science books and films (see also George Orwell's novels *1984* and *Animal Farm*). Two examples in this class are the 1964 *Summa Technologiae*, by the Polish writer Stanislas Lem, and *The Boys from Brazil*, by Ira Levin, published in 1976 (and turned into a film two years later by Franklin Schaffner, with Gregory Peck in the leading role). In 1923, English biologist John Haldane read his paper "Daedalus or Science and the Future" to the Heretics in Cambridge, and made it into a book in 1924. In this writing, which provided some inspiration for *Brave New World*, Haldane discussed ectogenesis and other biotechnology topics.

The theme of identical twins is present in Alexandre Dumas père's novel *The Man in the Iron Mask*, in which a French king keeps his identical twin captive and forced to wear an iron mask. Another relevant literary classic is the character of Faust, used by the English writer Christopher Marlowe and the German poet Johann von Goethe; this man acquires special powers over all other men by selling his soul to the devil, Mephistopheles. As early as 1818, the theme of an unscrupulous scientist creating a monster is described in the novel *Frankenstein*, by Mary Shelley, which forms the basis of many cinematographic adaptations. The myth of a man trying to challenge the creative powers of the gods goes back at least to Prometheus in Greek mythology. Greek mythology is full of other ideas about chimeras and combinational beings, half god and half human, or half human and half animal, such as the Minotaur, who had the body of a man and the head of a bull. Clones and identical twins are also related to the theme of a double, double life, double personality, the doppelgänger (German for double goer) found in many works of art and films.

Finally, the theme of human robots, humanlike machines, and other human replicas is abundant in the science fiction literature and films, such as the influential 1982 *Blade Runner*, with Harrison Ford. See also notes 4, 5, and 6 above.

## Chapter 6

1. In his book *Of Monsters and Prodigies*, the famous sixteenth-century surgeon Ambroise Paré proposed several explanations for this phenomenon, ranging from

the divine to the physiological, including: (1) God's demonstration of his power by sending things which are "unnatural"; (2) God's punishment for the mother's (or father's) wickedness; (3) intercourse during a woman's menstrual period; (4) slothfulness of the mother; and (5) an overabundance of sperm ejaculated during coitus (Guttmacher and Nichols, 1967).

2. The theme of Siamese twins is quite uncommon in our films and literature, with a few exceptions. Mark Twain seems to have been fascinated with Eng and Chang, and the subject of conjoined twins, as can be seen, for instance, in the appendix "Those Extraordinary Twins" in the novel *Pudd'nhead Wilson*. In the 1950s Vladimir Nabokov wrote a short story on Siamese twins titled "Scenes from the Life of a Double Monster." In 1969 John Barth wrote a short story, "Petition," centered on the conflict between Siamese twins. Judith Rossner's 1970s novel *Attachments* describes two women who marry conjoined twins. Siamese twins also play an important role in the 1995 surrealist French fairytale film *The City of Lost Children* by Jean-Pierre Jeunet and Marc Caro. *Joined at Birth* is the title of a documentary on the subject shown on the on the Discovery Channel cable network. Finally, "Sesame Street," a famous PBS series for children, has its own two-headed monster puppet.

3. In particular it is possible to alter or remove selectively any gene contained in stem cells' DNA. Modified stem cells are then injected into a mouse embryo. The resulting mouse is chimeric in the sense that a proportion of its organs and cells descends from the modified stem cells. The fundamental point is that some of the descendants of this mouse will be derived from gametes that contain the modified genes, and therefore will contain a copy of the modified gene in *all* their cells, not a fraction of them (like their parent). By breeding these descendants together, it is possible finally to obtain mice that are homozygotes for the altered genes; that is, both chromosomes in the corresponding pair carry the modified gene. This of course is invaluable to assess the properties of the gene, its function and dysfunction, and so forth. In particular, this model can be applied to genes that cause human diseases.

4. Even if they were symmetrical, this would lead to fairly different environments for each brain, notwithstanding the well-known differentiation that exists between the hemispheres, which could be exacerbated in Siamese twins.

5. Development can be manipulated in many ways, including surgically, chemically, and altering the substrate. Skin or bone cells, for instance, have been grown on artificial scaffolds.

## Chapter 7

1. Such a path is not easy to find because the evolutionary tree with its myriad random events is not directly available. Yet many of its properties can be progressively inferred by studying the genomic sequences of today's species, the fossil record, and so forth.

## Chapter 8

1. The Towers of Hanoi problem consists of three pegs and a number of rings of various sizes. Initially, the rings are stacked on the first peg in order of decreasing size, the smallest ring at the top. The goal is to move the rings from the first peg to the third peg one at a time, in such a way that no ring is ever placed over a smaller ring. The second peg may be used for temporary storage of rings.

2. Note the obvious consequence: the ultimate punishment for prison inmates, as well as unruly children, is "no Internet access." Of course, something similar could be said for computation. In an advanced democratic information society, all individuals may have an inalienable right to computing power.

3. To which the authors add: "The impact on humanity and the biosphere could be enormous, larger than the industrial revolution, nuclear weapons, or environmental pollution. We must take steps now to shape the emergence of artificial organisms; they have potential to be either the ugliest disaster, or the most beautiful creation, of humanity."

4. One could argue that running the simulations that generate artificial beings produces heat.

## Chapter 9

1. For a sense of the complexity associated with the total *number* of neurons, we can compare entire neurons to sand grains. We can roughly estimate the size of a sand grain to be on the order of 1 $mm^3$. This is an overestimate—think of it as very coarse sand. This corresponds to a blowup factor of 100, rather than the $10^7$ factor used earlier. Assuming the brain has $10^{12}$ neurons, $10^{12}$ grains of sand occupy a volume on the order of 1000 $m^3$, the size of a small beach: 1 m deep, 10 m wide, and 100 m in length!

To sense the complexity of neurons further, it is worthwhile to continue to zoom up to a factor of $10^7$ and again use the stadium analogy. At this scale, an axon, the main outgoing fiber of a neuron that carries its electric pulses and has a real diameter in the range of 1 $\mu$, looks like a fat pipe 10 m in diameter. The pipe can be quite long, branch, and extend all the way to other stadiums in the same city, in neighboring cities, and sometimes even to cities located on other continents (a 1 mm-long axon corresponds to a 10-km pipe). The dendritic tree of a neuron also forms a very extended structure compared with the stadium itself. Proteins (i.e., tennis balls), for instance, must be transported from the stadium to the tips of the axon. Certain proteins are embedded in the wall of the stadium and have the ability to allow for the local flow of certain grains of sand (ions) across the stadium wall. These channel proteins achieve a conduction rate of about $10^7$ ions per second, equivalent to several liters of sand per second. A net excess of 100 liters of a certain kind of sand (sodium ions, for instance) is sufficient to "depolarize" the stadium. When these local sand flows depolarize the stadium above a certain threshold, an electrical disturbance is produced—the action potential—that propagates along the fat pipe and reaches the synapses, where a signal is propagated to other neurons. Again we leave it as an exercise for the interested reader to take a closer look at the spatial and temporal scales involved with action potentials and synaptic transmission in a stadium-size neuron. It is also important to keep analogies in perspective and not get caught in them. After all, the physics of sand grains is different from the physics of ions in solution.

The complexity, and our related state of relative ignorance, extend from the level of molecules and single neurons to the level of entire brain regions and behavior. For large portions of the brain, even the rough function remains unknown. A good example is the cerebellum, a large, dorsally projecting part of the brain located between the brain stem and the back of the cerebrum. At the cellular level, the cerebellum has a repetitive, almost crystalline, architecture that has attracted the efforts of a number of brain modelers. For a long time, lesion studies have suggested that the cerebellum is involved mainly in motor control. More recent evidence, however, suggests that it is involved in many other activities, from sensory processing to learning, and clearly such multiplicity of functions need not be contradictory. In imaging experiments, for instance, the cerebellum "lights up" under many tasks that are not primarily motor tasks. Our lack of understanding of the cerebellum has even led some neuroscientists to suggest that the primary function of the cerebellum is to "compensate for its absence!"

2. Once I was collaborating with a neuroscientist on an olfactory experiment that involved teaching laboratory mice how to discriminate between odors. After many months of dedicated but frustrated attempts, we discovered by chance that the mice we had been working with had a number of genetic neurological defects; they lacked a major class of neurons in their olfactory systems as well as connections between their hemispheres! No wonder they could not learn—one cannot escape the law of the genes. In fact such deficiencies are common in populations of laboratory animals that have been inbred over many generations in order to preserve genetic uniformity. Deficiencies of this kind should always be considered carefully in behavioral experiments using these animals.

Curiously, the neuroscientist held the somewhat paradoxical and contrarian view that molecular biology, not neuroscience, was still in its infancy. What he meant is that molecular biology is in the process of studying the building blocks, DNA and genes. Neuroscience has already found the building blocks, that is, neurons, and is studying their complex interactions. And it is true that several different neuronal circuits are known in some detail in many different areas and organisms. On the other hand, very few gene regulatory circuits are understood entirely. Whereas in the end it is not very productive to pit one branch of biology against another, I completely disagree with this view for several reasons.

First of all, the number of building blocks is vastly different over several orders of magnitude: 40,000 genes versus 1 trillion neurons. Soon we will know all the genes individually, but we are far from knowing all the neurons individually. Second, the DNA sequence of a gene and its regulatory regions provides a fairly good, compact, and comprehensive description of each isolated building block. As pointed out, neurons are much fuzzier and more complex to describe. The connectivity of the circuits is also vastly different. A typical cortical neuron is connected to well over 10,000 other neurons. Detailed brain anatomy has not even been scratched, and even if it were available, we would not know what to do with it. At this level, gene regulatory circuits are definitely simpler. Finally, from an experimental standpoint, individual genes are much more accessible than individual neurons. In fact, over the next few decades we should be able to continue to gain invaluable insights about brain structure and function by continuing to study the brain's main molecules and proteins, its development, and the genes that control them.

3. The ancient Greeks had a prescientific theory of the atomic structure of matter. In fact, the word "atom" comes from the Greek *atomos* (indivisible).

4. I was once visiting a well-known neuroscientist who had such a patient. Luckily, that day the patient was due to visit the neuroscientist's office, and I was looking forward to meeting him, but he did not show up at the expected time. A driver, who probably had not been informed of his status, had brought him to the hospital from the convalescent home where he stayed, together with a number of other patients, and left them at the curb in front of the hospital! Clearly such a person does not really know who he is and has little sense of his own biography and his life. This patient, who had retained his natural jovial personality, could not function in society and hold any kind of job, but he could carry on a conversation and fool his interlocutor at least for a short time (in other words, he could pass a short Turing test). After several hours he was finally found where he had wandered almost randomly, happily conversing in the hospital's maternity ward.

5. Emotions seem to play an important role in guiding the decisions that are behind our actions. Our brains are faced with a world and an array of actions that are too complex to analyze in real time. The role of emotions (without explaining their underlying mechanism) could be to provide a computational shortcut. But if this is truly the main role for emotions, the logical consequence is that with very large computational power, the computation of the "best action" becomes possible in real time and emotions become useless. In other words, it is conceivable that emotions are necessary only for computing systems in a certain intermediate range. They are not required for the simplest and the most powerful of computing systems. The simplest systems process only a few bits, and confusion is hardly possible: single-cell organisms are unlikely to feel pain. The most computationally powerful systems can calculate the most rational action in real time without emotional shortcuts. In this context, the notion of computational power is relative to the complexity of the space of information and actions available to a given processing system. Therefore a converse equivalent statement of this hypothesis would be that if the space of information and available actions is simple with respect to the available computing power, emotions are not necessary. Beyond decisions, however, emotions seem to play an essential role in learning. As long as learning must go on throughout a lifetime, they cannot be discarded in a complex living organism. Thus an emotionless agent resembling a human being must have, as a minimum, the computational power required to (1) make comparable decisions at time $t$ and (2) continue to learn in a comparable fashion throughout a lifetime.

6. IBM has also produced ASCI White, the largest supercomputer in the world in 2000, the size of two basketball courts. The 2000 version of ASCI White has 8192 processors and is capable of 12.3 trillion operations per second. It is used by Lawrence Livermore National Laboratory to simulate nuclear test blasts that are banned by the 1996 Comprehensive Test Ban Treaty.

7. As we have seen, the fact that we will have available in silicon a raw number of operations per second comparable with the human brain does not imply we will be able to implement the same algorithms the brain uses, or to replicate human intelligence. First, we would have to know the algorithms. Second, it is likely that the algorithms are closely tuned to—in fact, almost inseparable from—brain wetware. In other words, they require neurons to run efficiently. But even today we have detailed models of single biological neurons implemented both in hardware and software. In the first case, the propagation of voltages and the generation of action potentials are mimicked by a collection of transistors. The artificial neuron is often much faster than its biological equivalent. In the second case, a neuron is simulated by an algorithm running on a computer. In complex cases, it may take several minutes to simulate the details of a few milliseconds of neuronal activity. Both types of models are still evolving, and we are only at the first stages, where networks, rather than isolated neurons, can be emulated. But clearly this is an area where plenty of progress could occur in the coming centuries.

8. This question has often been raised in various forms by linguists. Technically speaking, it is true that languages have the ability to produce an infinite number of sentences. Suffice it to think of syntactic structure such as "He said that she said that he said. . . ." Although I do not think that this is a central issue here, for all practical purposes we could restrict ourselves to a finite universe of possible sentences by putting a cap on the length of sequences. If a computer was capable of seamlessly conversing with a human on any topic using only sentences with fewer than 1000 words, I do not think Turing would object too much. Note that the same finite-infinite issue can be raised for chess.

9. This is another recurring theme in the science fiction literature. For a very mild version see, for instance, Michael Crichton's *The Terminal Man*, in which an external hardware feedback loop between different brain areas of an individual is created in an attempt to cure him of seizures. As an indication of things to come, in 1999 a conference sponsored by the National Institutes of Health, the National Institute of Mental Health, and the Alfred E. Mann Institute for Biomedical Engineering was held in Washington, D.C., with the title "Toward Replacement Parts for the Brain: Intracranial Implantation of Hardware Models of Neural Circuitry." The focus of the conference was on "one of the newest frontiers of neuroscientific and bioengineering research: the intracranial implantation of computer chip models of brain circuitry as neural prosthetics to replace damaged or dysfunctional brain tissue." Here and in many other areas covered in this book, nanotechnology—the development of tiny machines capable of operating on very small spatial scales, down to the molecular level—is bound to play an important role. Although still at an early stage, several molecule-size motors have been developed.

10. See, for instance, Mead 1989 and Maher et al. 1999. Several technologies for recording and imaging electrical, magnetic, and metabolic activity of brain areas

are available, such as EEG (electroencephalography), PET (positron emission tomography), and fMRI (functional magnetic resonance imaging). There are a few cases of blind patients for whom extremely primitive levels of vision have been restored by having visual information, captured from a television camera, processed by a computer and sent to the patient's visual cortex via a set of electrodes implanted in the brain. An example of such a wearable device has been produced by William Dobelle of the Dobelle Institute and Columbia-Presbyterian Medical Center in New York. Silicon microchips have been implanted beneath the retina of people suffering from blindness caused by retinitis pigmentosa, a hereditary degenerative disease. There is even the American Society of Artificial Internal Organs, which has its own journal and annual conference.

11. Other examples are artificial respirators and dialysis machines, which have been routinely used for quite some time.

12. We have seen that biotechnology can modify our sense of time in several ways—for instance, by reversing phylogenetic relationships. The same is true of computer technology with the exponential expansion in the amounts of information that can be stored, computed, and exchanged. The expansion in human and machine intelligence, and other faculties, has the potential for distorting our traditional notion of time by transiently accelerating civilization, the pace of discoveries and technological improvements, and, more generally, the speed at which information is processed and circulated. Many computationally driven events can happen much faster than the pace our brains have been accustomed to for the past 200,000 years.

13. We are skipping a number of nonessential technical details in this approximation and below. To be precise, a few other brain outputs affect the external world directly—for instance, through innervation of secretory glands such as lachrymal and sweat glands. Although we are mainly interested in computations performed by the brain, as a physical system the brain also participates in a number of exchanges with the outside world through the digestive, respiratory, and circulatory systems. Substances absorbed by the brain, whether drugs or food, affect its processing.

14. This estimate could be revised upward a little if, instead of binary spikes, we were interested in the actual noisy voltage samples, for instance, every millisecond. So the estimate could be as high as a few thousand bits per second, but this would not affect the sense of our argument very much.

15. See note 13. Additional second-order effects include the problem of recording inputs and outputs during the early development stages of the nervous system, and the fact that the anatomy of the input-output lines is variable. It varies both from individual to individual, and from time to time in a given individual, as a result of development, injury, degeneration, and so forth.

16. Most humans, including those interested in immortality, would feel very uncomfortable with the idea of a public permanent repository of complete external selves, let alone internal ones, especially in easily decodable audiovisual format, available to all for scrutiny. And this in spite of the fact that the external self encompasses pretty much all the history of an individual, its interactions with and impact on the outside world.

On the other hand, very fragmented, and possibly filtered, samples from the external self of each one of us are trivially available in the form of photos and audiovisual recordings. In fact, this digital approach to immortality is currently followed in several graveyards. Forever Enterprises in Saint Louis, for instance, has outfitted graves in seven cemeteries with "digital biographies" of the dead. By touching the screens of small video monitors housed in dark wooden kiosks, mourners can view photographs or videos of a deceased person, and listen to his or her voice. Brent Cassity, the CEO of the company, predicts that in the future, monitors will be attached to grave markers themselves. The company's 10,000-plus digital biographies are also available on the Internet. Leif Technologies in Ohio advertises funerary monuments equipped with personal memorials called "View★logies." View★logies can hold up to the equivalent of 250 pages of pictures and text, and can be downloaded to a laptop computer!

17. We could also try to develop brainless clones, although it is unlikely that such practice would qualify as being more ethical.

18. See Griffith et al. 1999.

19. The most extreme form of molecular medicine is the idea of "molecular doctors"—small molecular assemblies capable of making a diagnosis and taking remedial actions. Modern drug design and molecular computing technology are still very far from that. Yet, to some extent, we are beginning to be able to build molecules that can recognize or bind to other molecules, or molecules that can compute the solution of simple problems via their chemical interactions, as in the field of DNA computing.

20. See Lowenstein and Parent 1999; Tropepe et al. 2000.

21. See Eriksson et al. 1998; Gould et al. 1999.

22. Injections of neural stem cells are capable of restoring normal function in laboratory mice (Yandava et al. 1999). Neural stem cells can develop into neurons, glial cells, and other cells in the central nervous system. Neural stem cells from culture were injected into the brains of newborn mice from a strain called "shiverer." Shiverer mice have a genetic defect and cannot produce a protein that is required to form myelin, the fatty tissue that surrounds and insulate axons, thereby greatly increasing their electrical conductivity. As a result, these mice develop tremors in the first three weeks of life, and subsequently other symptoms similar to those seen in humans with multiple sclerosis (the disease is characterized by a progressive loss of myelin). Sixty percent of the injected mice appeared to behave like normal mice. Examination of the animal's brains provided evidence that the injected stem cells are capable of migrating, detecting the problem, and converting into brain cells that produce myelin (oligodendrocytes). Similar experiments on adult mice are in progress. Related results can be found in the article by Brüstle and collaborators (1999).

Together with other experiments, this provides initial evidence that neural stem cells can migrate to where they are needed, and possibly replace damaged or missing cells in the brain. Injection of stem cells could become part of therapy for a wide range of acute and chronic neurological disorders, including degenerative disorders such as Alzheimer's and "mad cow" diseases, strokes, multiple sclerosis, injuries, and many brain birth defects.

23. All forms of large-scale transplants are bound to create some discontinuity and disruptions in how information is processed in the central nervous system. It is clear that transplanting an eye in humans is several orders of magnitude more complex than transplanting a liver, both in terms of keeping all the cells alive and especially in terms of reconnecting the optic nerve with its 1 million axons. Even in case of successful reconnection, a very difficult period of relearning to see would necessarily ensue, requiring visual areas in the brain to recalibrate themselves to the new input. Under natural conditions, such plasticity exists but decreases in scope during development and with age. Future technology should learn how to manipulate such plasticity. To make matters even more complex, the eye is not a purely sensory structure. A difficult problem is also posed by trying to reestablish the delicate control of eye movements. And whereas the transplant of an eye is at least conceivable because the eye is at the periphery of the nervous system, transplantation of modules that are buried deep inside the brain is extremely difficult to imagine.

Something else to consider is transplantion of entire heads instead of components. Conceptually this is not completely dissimilar from transplanting an eye (from a purely technical standpoint, heads might even be easier), the main problem here being reattachment of nerve fibers in the spine and subsequent relearning. All these transplant operations are of course helped by cloning technology, which in principle can provide compatible tissues and organs of any age and developmental stage. But even if transplanting heads became possible one day, by itself it would not solve the problem of brain mortality because at best it would provide a brand-new body to an older brain. Note, however, that such technology would also open the door for transplanting heads onto genetically different bodies, as long as techniques are available for controlling rejection. That would further expose the intrinsic continuity found at the levels of genomes and of brains.

A final technique worth mentioning in the quest for immortality is freezing. At low temperatures, all metabolic reactions are slowed down or halted. It is therefore conceivable that bodies and brains could be frozen and preserved, then revived at a later date. "Cryonics" is a recently coined word referring to the practice of preserving the whole body, head, or brain of persons declared legally dead, in the hope of revival at some time in the future. Long-term storage at low temperatures uses immersion in liquid nitrogen. The American Cryonics Society (ACS), founded in 1969, promotes this possibility, and a number of companies already provide an array of cryonic services. Preservation of entire body, head, and brain is available for typical respective prices of $140,000, $70,000, and $30,000. Interest on the funds remaining after the initial suspension costs must be sufficient to cover annual liquid nitrogen costs. An example of a typical cost breakdown provided by the company Alcor, rounded to the nearest fifty dollars, is given in the table below. Additional services provided by the ACS include preservation of a DNA sample at room temperature in a glass vial, at a cost of $50, through an arrangement with Third Millennium Research, Inc. of Seattle. Available for $150 is indefinite cryogenic storage of a biological tissue sample.

Typical cost breakdown for cryonics services provided by Alcor

| | Whole Body ($) | Neuro ($) | |
|---|---|---|---|
| Remote transport | 14,050 | 14,050 | |
| Cryoprotective perfusion | 13,400 | 11,500 | |
| Laboratory evaluations | 950 | 950 | |
| Temperature descent | 8,350 | 1,750 | |
| Record keeping | 450 | 400 | |
| Total | 37,200 | 28,650 | |
| Annual liquid nitrogen | 850 | 50 | "Bigfoot" dewar |
| Storage costs | 1,700 | 150 | Older-style dewar |

It should not be thought, however, that freezing is just a form of immobilization that occurs without deep alterations in, and even damage to, biological components. In particular, extensive research on the "reviving" step of cryonics remains to be done. Yet in principle, nothing suggests that such a difficult step may not become possible with scientific and technological progress. We are able to revive people who have drowned in freezing waters after relatively long periods of time. In addition, frozen sperm or embryos are routinely used in assisted reproductive techniques.

But even with advanced cryonic technology, freezing by itself creates immortality only for a frozen body. One could conceivably alternate between periods of frozen and nonfrozen states spread over a very long time, such as several centuries. Though this raises a host of interesting issues, it does not by itself guarantee eternal life: only an arbitrarily long but intermittent one, assuming that the aging process is roughly turned on and off in the corresponding phases.

In passing, it is also worth noting that lowering the temperature of body parts, including the brain, to reduce metabolic damage and cell death is currently performed, for instance, after a stroke.

24. IP stands for Internet protocol. The IP address is the characteristic address each computer has on the Internet. Current addresses consist of twelve decimal numbers partitioned into four groups of three. Note also that even at the computational level, two computers of the same brand and model are extremely similar but not perfectly identical. For instance, they may fail at different times, have slightly different memory sizes, or take minutely different amounts of time to perform the same computation.

25. In a mathematical sense, a distance over a set is a function that associates a positive number d(x,y) to each pair (x,y) of elements in the set. In addition, it must satisfy three requirements:

$d(x,y) = 0$ if and only if $x = y$;

$d(x,y) = d(y,x)$ (symmetry);

$d(x,y) + d(y,z) \geq d(x,z)$ (triangle inequality).

The third requirement is the most constraining.

26. The distance between two brain anatomies or physiologies is more difficult to calculate than the distance between two genomes. But it is clear that one could start with gross approximations based, for instance, on the size, weight, and activation patterns of macroscopic areas and structures, which can be obtained in part by imaging techniques. Technological refinements over several centuries could lead to more and more precise measures. But in many ways the genomic and anatomical components of the distance are not the most interesting ones. The computational component of the distance is the most relevant.

27. An issue that is important, but not essential for our perspective, is whether having similar external selves implies having the same subjective sense of self. As long as the notion of subjective sense of self remains undefined, it is obviously difficult to address this question, but intuitively it would seem that the answer is no, although that would not really matter for the external world. A related issue, partially discussed in appendix E, is how much the external self constrains the internal self.

28. It should come as no surprise that the general learning problem is NP complete, and therefore likely to be intractable in its generality. But this is just a starting point rather than a final negative conclusion. Plenty of work remains to be done with the space of learnable functions and their approximations.

29. Instead of being identical, a C sharp and the consecutive D flat, for instance, were separated by 1/9th of a tone, the C sharp being located 4/9th of a tone higher than the C, and the D flat being 4/9th of a tone lower than the D.

30. We could also work on the acoustic environment, on the dryness of the air, and more generally on the medium through which sound is propagated, but this approach has a relatively limited scope.

31. There are technical problems, such as synchronization, but in principle they can be overcome.

32. Curiously, one can associate sounds with DNA sequences of letters and then "play" an entire genome. It remains to be seen whether an encoding exists that would produce interesting music in this way. But conceivably, encodings could exist so that genetic patterns such as gene regions could become detectable by human ears, which are endowed with great pattern recognition abilities.

33. This is true of many other human activities. The auditory system was not *designed* by nature, and certainly not designed to listen to music. Initially, it evolved by evolutionary tinkering under other selective pressures, such as recognizing the sounds of predators and other important elements in the immediate environment, and probably human communication. Thus, music would appear to be a by-product. But then how can it be so powerful? A few notes can completely alter our mood or change our motor state, therefore producing massive changes in brain electrical and chemical activity. Animals also show some reaction to music and simple rhythms. The philosopher Leibniz wrote, "The pleasure we obtain from music comes from counting, but counting unconsciously. Music is nothing but unconscious arithmetic." The connection between music and numbers is interesting and has been made by many others, but it hardly provides an explanation for

why we derive pleasure from music. The strength of our attachment to and need for music suggests to me that there must be something physiological about it that attracts us, a little like a drug. Music must produce patterns of neuronal activity and biochemical alterations that in the end are pleasurable or addictive, a "massage" of certain neuronal circuits.

## Chapter 10

1. It is common in IVF to create several embryos at once in a dish and implant several of them in the future mother. The remaining embryos are usually frozen for future use and typically are discarded if not used after a certain time. These are the embryos from which embryonic stem cells can be derived. The main reason for implanting several embryos in the mother is to increase the chances of a successful implant that results in a well-established pregnancy, rather than the production of multiple births. When more than one implanted embryo survives, the mother is offered a selective abortion in which unwanted embryos are removed.

A different ethical side issue related to embryos and reproductive genetics is selection of sex. The ethics committee of the American Society for Reproductive Medicine stated that selecting embryos based on sex in assisted reproduction procedures is not acceptable unless it is done to avoid genetic diseases. The committee did not recommend that sex selection for nonmedical reasons be made illegal but, warned that such practices "give cause for serious ethical caution."

A legal case reported by the *Boston Globe* concerns the fate of four frozen human embryos, locked away in a fertility clinic, that were produced by a couple who are now divorced. The woman wanted to carry at least one of the embryos to term in order to expand her current family, but her ex-husband did not agree. What is the legal status of human embryos? Who owns them? Other cases of embryo custody involving divorces have occurred. In 1992 the Tennessee Supreme Court ruled, in a case where there was no written agreement between the parties, that the man could not be forced into fatherhood. In 1998 New York's Court of Appeals ruled that five similarly contested frozen embryos ought to be donated to research and ultimately destroyed.

2. The difficulties are accentuated by the fact that at the cellular level we again find the problem of continuity at the level of entire organisms. Separation between cell types (pluripotent, totipotent, tissue, and organ cells) is probably not absolute, and in time we may learn to evolve from any one type to any other through a series of small, reversible changes. If this were the case, it is difficult to find an absolute principle that would allow manipulation or even destruction of certain cell types, say skin cells, but forbid the same operation on other cell types.

3. In my opinion, another murky area where our laws, customs, and thinking will have to be overhauled concerns sex-marriage-reproduction.

4. The U.S. Patent Office has progressively moved from what it calls "generation one" patents, based on sequence alone, to "generation two" patents, based on sequence and some functional insight derived from homology considerations (similarity to known genes), to "generation three" patents that include laboratory infor-

mation about the gene's function, biochemistry, cell biology, or genetics. More generally, for hierarchically organized information systems, biological as well as software, it may be useful to introduce general rules preventing patents on low-level information to dominate higher levels of the information hierarchy.

5. A similar problem may be occurring with gene fragments called ESTs (expressed sequence tags) that for a while were patentable. Because different ESTs can originate from one gene, different patent holders may share a stake in the same gene.

6. Note that the information storage capacity of DNA could be used to embed time bombs or watermarks inside genomes, in the same way this is done for software of image data.

7. In January 2000, the National Institutes of Health (NIH) announced a new policy allowing NIH-funded scientists to use oncomice and other patented transgenic animals for noncommercial research, without obtaining specific approval from Du Pont. In particular, Du Pont has agreed not to seek broad rights to discoveries made in nonprofit laboratories through research based on patented animals.

8. Decision theory is a branch of science originally developed in the first half of the twentieth century by mathematicians, such as John von Neumann, aiming at modeling the process by which optimal rational decisions are made. In a nutshell, imagine a decision maker faced with an uncertain environment, a choice of possible actions, and a system of costs or rewards associated with each action and each state of the environment. Under a small set of reasonable axioms, the optimal strategy ought to maximize the expected gain or minimize the expected loss. The notion of expectation refers to a weighted average taken over all possible states of the environment. Thus, decision theory naturally leads to probability theory as a necessary intermediary step to define such weights or probabilities. A generalization of decision theory is game theory, in which several decision makers compete, form alliances, and so forth. In spite of its name, one of the real intents of game theory is to model complex socioeconomic situations.

9. To a first approximation, the utilitarian-Kantian divide is reflected in the opposition between countries (United States and United Kingdom versus other European countries) and religions (Protestantism, Judaism, and even Islam versus Roman Catholicism). Germany, for instance, where biological concepts such as selection still resonate with the specter of Nazi demagogy, has very strict rules against human cloning. In Italy, both human and animal cloning have been forbidden by law. The same day the article on Dolly was published, the Roman Catholic Church declared human cloning to be unacceptable in its official newspaper, *L'Osservatore Romano*. In 1998 the parliament of the European Union voted an official document forbidding cloning, and more generally the artificial creation of genetically identical humans, adults or embryos, by any means. On the other hand, the initial impact of Dolly in the United States and the United Kingdom was followed by much milder measures, such as a ban on publicly funded research on cloning, leaving the door open for privately funded operations or a five-year moratorium with the possibility of reevaluating the matter in a flexible way in the near future. As a side issue, it is worth noting the related contemporary divide between the United States

and most of Europe regarding genetically modified (GM) crops and foods, derisively called "frankenfood" in Europe. The gap was only slightly reduced by the decision in the United States to disallow the label "organic" on GM foods.

10. In my opinion, this is mostly a rhetorical question. To a very large extent, it would seem that Hitler was the product of particular historical circumstances, and it is doubtful that his clone could lead current Western societies toward another holocaust. On the other hand, the broader issue of whether criminal behavior can have a genetic component, and, if so, whether such a component should be manipulated, is a legitimate one.

11. In the letter of transmittal to the president, the Commission stated:

*In this report, we address a very specific aspect of cloning namely where genetic material would be transferred from the nucleus of a somatic cell of an existing human being to an enucleated human egg with the intention of creating a child. We do not revisit either the question of the cloning of humans by embryo-splitting or the issues surrounding embryo research. The latter issue has, of course, recently received careful attention by a National Institutes of Health panel, the Administration, and Congress.*

*Not surprisingly, we have discovered that the potential ability to clone human beings through the somatic cell nuclear transfer techniques raises a whole host of complex and difficult scientific, religious, legal and ethical issues—both new and old. Indeed, the Commission itself is unable to agree at this time on all the ethical issues that surround the issue of cloning human beings in this manner. It seems clear to all of us, however, given the current stage of science in this area, that any attempt to clone human beings via somatic cell nuclear transfer techniques is uncertain in its prospects, is unacceptably dangerous to the fetus and, therefore, morally unacceptable. At present, moral consensus on this issue should be easily achieved. Furthermore, the continuing controversy over the social and ethical issues raised by this new development requires more time for deliberation and the accumulation of new scientific data. We therefore recommend that the current moratorium on attempts to create children in this manner be continued and that you immediately ask for voluntary compliance in the private sector while federal legislation banning the use of these techniques for creating children is formulated and considered.*

12. Animal cloning can be useful in many ways: to save endangered species, to improve selection, to study the effects of nature versus nurture, and to reduce the number of experiments and animals required to test a drug or a new therapy by allowing a tight control of genetic variables. It can also be effective when it is desirable to make copies of a genetically engineered organism and sexual reproduction is impossible, inefficient, or not desirable.

13. The fact that new procedures are initially expensive and available only to affluent people can be viewed as unjust. On the other hand, it provides an inertia mechanism that results in slow diffusion, and this could be beneficial, inasmuch as we want to be prudent with the use of new technologies.

14. See Solter and Gearhart 1999. Often, fertility techniques can propagate fertility problems to the next generation. In this case, the son or daughter may not necessarily be infertile if the infertility of the corresponding parent is genetic and is confined to one dominant gene not present in the offspring.

15. Some contend that cloning stands a chance of violating the basic laws of democracy and dignity because some individuals are created not equal, and may disrupt families and the social fabric with its genealogical lineages by the time-warping effects discussed in a previous chapter. It can pervert "normal" relationships, for instance, by having identical twins with different mothers, fathers and sons of the same age, and so on. All these dangers are greatly exaggerated and are present in other technologies in use today.

16. *Science* Vol. 283, No. 5409, 19 March 1999, p. 1849.

17. As another sign of rapid pace in this area, the University of Winsconsin has created the WiCell Research Institute to distribute stem cells to academic and industrial scientists, starting in the spring of 2000. Thomson is the scientific director of the Institute. In Japan, a committee of the Council for Science and Technology, the nation's highest science policy body, has endorsed the use of human stem cells in research. It has gone even further by recommending government funding for both the derivation and the use of stem cells. Advertisements for fetal or embryonic tissue are commonly found in scientific journals such as *Science*.

18. The U.S. Congress is sharply divided on these issues. A bill has already been proposed by Senator Arlen Specter (R., Pennsylvania) to allow federal funding for the use and *derivation* of embryonic stem cells. In Australia, it is currently illegal to take cells from human embryos, but it is legal for the cells to be obtained from another country and then used in Australia.

19. This is not to say that during the transitory phase we live in, the *process* or *intent* by which a cell is manipulated may be regulated. But this is different from relying on some metaphysical property of the embryo.

20. Duchenne muscular dystrophy is one of the most common and most devastating human genetic disorders, affecting all populations worldwide. Clinically, it is considered the most important neuromuscular disorder. Named after the French physician who first described the condition, it is usually fatal and consists of a progressive weakening and degeneration of the voluntary muscles that control movement. Symptoms usually appear between the ages of two and six years. It is estimated that 1 in every 4000 newborn boys in the United States has Duchenne muscular dystrophy. The gene responsible for the disorder was identified in 1986 and codes for a protein called dystrophin. Without dystrophin, muscle fibers degenerate and cannot work properly. The gene is located on the X chromosome, which explains why the disease is found almost exclusively in boys. Males are characterized by a pair of XY sexual chromosomes. With a single X chromosome, any abnormality of the dystrophin gene will cause muscular dystrophy. Females, on the other hand, are characterized by a pair of XX chromosomes. If one copy of the gene is defective, usually a normal copy of the gene on the other X chromosome provides adequate production of the protein. This is a complex disorder, and many hurdles stand in the way of developing treatment based on normal gene delivery.

21. Consider, for example, the argument that genetically infertile parents should not be allowed to clone themselves on the ground that their clones would also be

infertile. It is clear that the day corresponding gene therapy becomes available, this argument becomes entirely moot.

A second example has to do with the fear that by modifying genes and not letting natural selection run its course, we risk compromising the human genetic pool. Imagine, for the sake of argument, that we learn how to fix gene A when it is defective, and that defects in gene A correlate with defects in gene B. Assume also that we do not know how to fix gene B. Could the defective gene B spread in the population because the survival rate of individuals with defects in A and B is increased by the gene A therapy? It is quite unlikely in reality (and the risk can be estimated) because therapy for gene A will initially be expensive and concern only an extremely small fraction of the total population. It will also take several decades for cured individuals to reach adulthood and reproduce. By that time, we are likely to have found a therapy for gene B, too. In other words, the speed at which scientific knowledge is progressing, combined with our increasing control of reproduction, is likely to exceed the speed at which large-scale genetic catastrophes could propagate in the human population by natural (and obsolete) means.

22. Like any technology, the Internet is not good or bad per se. Many opposite dynamic forces coexist around it that can affect relationships in many ways. And their outcome is highly unpredictable. The Internet, for instance, is being extensively used for matchmaking purposes. It also allows for deployment of completely new strategies between the sexes. Some women, for example, have been quick to cash in on men's interest in cybersex by opening some of the most visited cybersex sites. Although these sites are still a small minority, one cannot exclude the possibility that one day we will evolve toward a situation with one cybersex site per human, and where computers will be only one of the forms silicon takes in the bedroom. It is remarkable to see how the relationship and sex industries have always been at the business and rapid adoption forefronts of new media technologies, from print to wireless.

Other examples of unusual strategies, or reinvention of old ones, include sites for the detection of cyber infidelities (www.infidelitybusters.com), the arrangement of infidelity alibis (www.alibi.co.uk), and new matchmaking strategies. Skim.com's vision (www.skim.com), for instance, is to integrate online communication with street communication worldwide by tagging clothes, artwork, and other products with numbered codes. These numbered codes are E-mail addresses on skim.com that can be used to get in touch and communicate.

23. Loosely speaking, a dynamic system is said to be chaotic if very small perturbations at a given time result in trajectories that diverge very rapidly (exponentially fast) in the future. Hence a chaotic system is extremely sensitive and unpredictable, at least in its finest details.

24. The growing disparity in scientific and technical knowledge among humans should definitely be on the list of important concerns. Not only will such disparity be reflected in the differences in wages, wealth, and power among different groups, it could also prevent the dialogue necessary to reach consensus on issues such as the use of embryonic stem cells or bioengineered organisms. Considerable knowledge oddities, obscuring the minds of large numbers of humans, continue to be found,

even in the most technically sophisticated societies. After all, in August 1999, the Kansas State Board of Education voted to delete evolution, as well as cosmology, from the mandatory science curriculum. Such an irrational and backward decision is baffling. As Stephen Jay Gould puts it: "What strange set of historical circumstances, what odd disconnect between science and society, can explain the paradox that organic evolution—the central operating concept of an entire discipline and one of the firmest facts ever validated by science—remains such a focus of controversy, even of widespread disbelief, in contemporary America?" More recently, an amendment to a bill that the Oklahoma legislature passed on April 5, 2000, by a vote of 99 to 0 reads: "When adopting science textbooks, the Committee shall ensure that the textbooks include acknowledgment that human life was created by one God of the Universe." Obviously the authors of such bill never made the effort to take a close look at the fossil record and the DNA of humans and chimpanzees. The formidable expansion in scientific knowledge could also pose a burden for the new generations that must assimilate it. In the meantime, public support remains essential to ensure adequate funding for the scientific enterprise, and can only be weakened by a wide disconnect.

## Chapter 11

1. And if you need to visit a bank in person, make sure you do it before going to the Irvine Spectrum mall, for none is anywhere in sight there. Like any respectable mall, Irvine Spectrum lies close to the intersection of two of the busiest freeways in the world . . .

2. Jules Verne wrote such popular titles as *20,000 Leagues Under the Sea, Around the World in 80 Days, From the Earth to the Moon*, and *Journey to the Center of the Earth*. The first circumnavigation of Earth in a hot air balloon took place in 1999.

3. The desire for immortality is a relic of the old self-centric view of the world, and may change considerably, or even vanish, as the boundary between life and death becomes fuzzier and fuzzier.

## Appendix E

1. Out of the 40,000 genes we have, only a fraction—on the order of 10%—may correspond to proteins that are relevant for brain computations. On the other hand, many proteins are present in multiple, slightly different variants (in the range of 10 to 100).

2. Recently, S. Lloyd has argued that the "ultimate laptop"—with a mass of 1 kg, occupying a volume of 1 liter, and operating at the limits of speed and memory space allowed by the law of physics—can perform $10^{51}$ operations per second on $10^{31}$ bits. In the unlikely event Moore's law were to remain true in the coming centuries, and given that today's computers operate roughly at $10^{10}$ operations per second on $10^{10}$ bits, it would take us only 250 years to make up this gap of 40 orders of magnitude.

# Bibliography

I have listed first a number of general scientific references and review articles, together with literary references, followed by a list of more technical articles for the reader interested in studying these topics in greater depth. Neither list is meant to be exhaustive; they merely provide background material and pointers toward the rest of the literature.

## General References

Abelson, P. H. (2000). "Funding the Nanotech Frontier." *Science* 288: 269.

Abelson, P. H., and Hines, P. J. (1999). "The Plant Revolution." *Science* 285: 367–368.

Annas, G. J., Caplan, A., and Elias, S. (1999). "Stem Cell Politics, Ethics, and Medical Progress." *Nature Medicine* 5: 1339–1341.

Banfield, J. F., and Marshall, C. R. (2000). "Genomics and the Geosciences." *Science* 287: 605–606.

Barinaga, M. (1999). "New Clues to How Neurons Strengthen Their Connections." *Science* 284: 1755–1757.

Barinaga, M. (2000a). "Fetal Neuron Grafts Pave the Way for Stem Cell Therapies." *Science* 287: 1421–1422.

Barinaga, M. (2000b). "Two Feedback Loops Run Mammalian Clock." *Science* 288: 943–944.

Barth, J. (1969). *Lost in the Funhouse.* New York: Bantam Books.

Barton, J. H. (2000). "Reforming the Patent System." *Science* 287: 1933–1934.

Berg, P., and Singer, M. (1998). "Regulating Human Cloning." *Science* 282: 413.

Borges, J. L. (1993). *Ficciones.* New York: Everyman's Library.

Brenner, S. (2000). "The End of the Beginning." *Science* 287: 2173–2174.

Buss, D. M. (1994). *The Evolution of Desire.* New York: HarperCollins.

Butler, D., and Smaglik, P. (2000). "Draft Data Leave Geneticists with a Mountain Still to Climb." *Nature* 405: 984–985.

Campisi, J. (2000). "Aging, Chromatin, and Food Restriction—Connecting the Dots." *Science* 289: 2062–2063.

Cavazzana-Calvo, M., Hacein-Bey, S., Saint Basile, G. de, Gross, F., Yvon, E., Nusbaum, P., Selz, F., Hue, C., Certain, S., Casanova, J., Bousso, P., Le Deist, F., and Fischer, A. (2000). "Gene Therapy of Human Severe Combined Immunodeficiency (SCID)-X1 Disease." *Science* 288: 669–672.

Cho, M. K., Caplan, A. L., McGee, D., and the Ethics of Genomics Group. (1999). "Ethical Considerations in Synthesizing a Minimal Genome." *Science* 286: 2087–2090.

Collins, D. (1994). *Eng & Chang: The Original Siamese Twins.* Morristown, N.J.: Silver Burdett.

Collins, F. S., Patrinos, A., Jordan, E., Chakravarti, A., Gesteland, R., Walters, L., and the members of the DOE and NIH planning groups. (1998). "New Goals for the U.S. Human Genome Project: 1998–2003." *Science* 282: 682–689.

David Smith, J. (1988). *Psychological Profiles of Conjoined Twins.* New York: Praeger.

Davis, P. (1999). "Nanotechnology: Synthetic Molecular Motors." *Nature* 401: 120–122

Dawkins, R. (1989). *The Selfish Gene.* 3rd ed. Oxford, UK: Oxford University Press.

DeSalle, R., and Lindley, D. (1998). *The Science of Jurassic Park and the Lost World.* New York: HarperPerennial.

Dickson, D. (2000a). "NIH Opposes Plans for Patenting 'Similar' Gene Sequences." *Nature* 405: 3.

Dickson, D. (2000b). "The Story So Far. . . ." *Nature* 405: 983–985.

Djerassi, C. (1999). "Sex in an Age of Mechanical Reproduction." *Science* 285: 53–54.

Drews, J. (2000). "Drug Discovery: A Historical Perspective." *Science* 287: 1960–1964.

Enserink, M. (2000a). "Patent Office May Raise the Bar on Gene Claims." *Science* 287: 1196–1197.

Enserink, M. (2000b). "Are Placebo-Controlled Drug Trials Ethical?" *Science* 288: 416.

Ferber, D. (1999a). "Lab-Grown Organs Begin to Take Shape." *Science* 284: 422–425.

Ferber, D. (1999b). "GM Crops in the Cross Hairs." *Science* 286: 1662–1666.

Ferber, D. (1999c). "Growing Human Corneas in the Lab." *Science* 286: 2051–2053.

Finkel, E. (2000). "Australian 'Ranch' Gears Up to Mass-Produce Mutant Mice." *Science* 288: 1572–1573.

Foster, K. R., Vecchia, P., and Repacholi, M. H. (2000). "Science and the Precautionary Principle." *Science* 288: 979–981.

French Anderson, W. (2000). "The Best of Times, the Worst of Times." *Science* 288: 627–629.

Gage, F. H. (2000). "Mammalian Neural Stem Cells." *Science* 287: 1433–1438.

Gaskell, G., Bauer, M. W., Durant, J., and Allum, N. C. (1999). "Worlds Apart? The Reception of Genetically Modified Foods in Europe and the US." *Science* 285: 384–387.

Gibbons, A. (2000). "Building a Case for Sequencing the Chimp." *Science* 289: 1267.

Gibbs, J. B. (2000). "Mechanism-Based Target Identification and Drug Discovery in Cancer Research." *Science* 287: 1969–1973.

Gilbert, S. F. (1997). *Developmental Biology*. 5th ed. Sunderland, Mass.: Sinauer.

Gould, S. J. (1999). "Darwin's More Stately Mansion." *Science* 284: 2087.

Guerinot, M. L. (2000). "The Green Revolution Strikes Gold." *Science* 287: 241–243.

Gura, T. (1999a). "Repairing the Genome's Spelling Mistakes." *Science* 285: 316–318.

Gura, T. (1999b). "New Genes Boost Rice Nutrients." *Science* 285: 994–995.

Gura, T. (2000c). "Reaping the Plant Gene Harvest." *Science* 287: 412–414.

Guttmacher, A., and Nichols, B. (1967). "Teratology of Conjoined Twins." In: *Conjoined Twins,* Bergsma, D., editor. New York: The National Foundation–March of Dimes.

Harris, J. (2000). "Intimations of Immortality." *Science* 288: 59.

Hagmann, M. (2000a). "New Genetic Tricks to Rejuvenate Ailing Livers." *Science* 287: 1185–1187.

Hagmann, M. (2000b) "Protest Leads Europeans to Confess Patent Error." *Science* 287: 1567–1569.

Herrmann, B., and Hummel, S. (1994). *Ancient DNA*. Berlin: Springer-Verlag.

Hillis, D. M. (1999). "Predictive Evolution." *Science* 286: 1866–1867.

Huxley, A. (1998). *Brave New World*. New York: HarperPerennial.

International Human Genome Sequencing Consortium (2001). "Initial Sequencing and Analysis of the Human Genome." *Nature* 409: 860–921.

International SNP Map Working Group (2001). "A Map of Human Genome Sequence Variation Containing 1.42 Million Single Nucleotide Polymorphisms." *Nature* 409: 928–933.

Irion, R. (2000). "The Science of Astrobiology Takes Shape." *Science* 288: 603–605.

Kaiser, J. (2000). "From Genome to Functional Genomics." *Science* 288: 1715.

Kandel, E. K., Schwartz, J. H., and Jessell, T. M. (1991). *Principles of Neural Science.* 3rd ed. New York: Elsevier.

Kennedy, D. (2000). "Two Cheers for New Stem Cell Rules." *Science* 289: 1469.

Knoppers, M., Hirtle, M., and Glass, K. C. (1999). "Commercialization of Genetic Research and Public Policy." *Science* 286: 2277–2278.

Kuhn, T. S. (1996). *The Structure of Scientific Revolutions.* 3rd ed. Chicago: University of Chicago Press.

Lander, E. S., and Weinberg, R. A. (2000). "Genomics: Journey to the Center of Biology." *Science* 287: 1777–1782.

Ledeberg, J. (2000). "Infectious History." *Science* 288: 287–293.

Lem, S. (1964). *Summa Technologiae.* Published in German as *Suhrkamp Tasch Enbuch Verlag.* Verlag Pocket Book, 1981.

Lenoir, N. (2000). "Europe Confronts the Embryonic Stem Cell Research Challenge." *Science* 287: 1424–1426.

Levin, I. (1976). *The Boys from Brazil.* New York: Bantam Books.

Lovett, R. A. (2000). "Toxicologists Brace for Genomics Revolution." *Science* 289: 536–537.

Lucky, R. (2000). "The Quickening of Science Communication." *Science* 289: 259–264.

Macilwain, C. (2000). "World Leaders Heap Praise on Human Genome Landmark." *Nature* 405: 983–984.

Mahoney, R. J. (2000). "Opportunity for Agricultural Biotechnology." *Science* 288: 615.

Malakoff, D. (2000). "The Rise of the Mouse, Biomedicine's Model Mammal." *Science* 288: 248–253.

Marshall, E. (1998). "A Vesatile Cell Line Raises Scientific Hopes, Legal Questions." *Science* 282: 1014–1015.

Marshall, E. (1999a). "Sequencers Endorse Plan for a Draft in 1 Year." *Science* 284: 1439–1441.

Marshall, E. (1999b). "A High-Stakes Gamble on Genome Sequencing." *Science* 284: 1906–1909.

Marshall, E. (1999c). "Legal Fight over Patents on Life." *Science* 284: 2067–2068.

Marshall, E. (1999d). "Ethicists Back Stem Cell Research, White House Treads Cautiously." *Science* 285: 502–503.

Marshall, E. (1999e). "Gene Therapy Death Prompts Review of Adenovirus Vector." *Science* 286: 2244–2245.

Marshall, E. (2000a). "FDA Halts Gene Therapy Trials at Penn." *Science* 287: 565–567.

Marshall, E. (2000b). "NIH Cuts Deal on Use of Oncomouse." *Science* 287: 567.

Marshall, E. (2000c). "Patent on HIV Receptor Provokes an Outcry." *Science* 287: 1375–1377.

Marshall, E. (2000d). "The Business of Stem Cells." *Science* 287: 1419–1421.

Marshall, E. (2000e). "Talks of Public-Private Deal End in Acrimony." *Science* 287: 1723–1725.

Marshall, E. (2000f). "Clinton and Blair Back Rapid Release of Data." *Science* 287: 1903.

Marshall, E. (2000g). "Claim and Counterclaim on the Human Genome." *Science* 288: 242–243.

Marshall, E. (2000h). "A Deluge of Patents Creates Legal Hassles for Research." *Science* 288: 255–2257.

Marshall, E. (2000i). "Gene Therapy on Trial." *Science* 288: 951–957.

Marshall, E. (2000j). "Rival Genome Sequencers Celebrate a Milestone Together." *Science* 288: 2294–2295.

Marshall, E. (2000k). "Moratorium Urged on Germ Line Gene Therapy." *Science* 289: 2023.

Marshall, E., Pennisi, E., and Roberts, L. (2000). "In the Crossfire: Collins on Genomes, Patents, and 'Rivalry.'" *Science* 287: 2396–2398.

Marx, J. (2000). "DNA Arrays Reveal Cancer in Its Many Forms." *Science* 289: 1670–1672.

McLaren, A. (2000). "Cloning: Pathways to a Pluripotent Future." *Science* 288: 1775–1780.

Moffat, A. S. (2000). "Transposons Help Sculpt a Dynamic Genome." *Science* 289: 1455–1457.

Morel, V. (2000). "Revealing a Dinosaur's Heart of Stone." *Science* 288: 416–417.

Nabokov, V. (1958). *Nabokov's Dozen.* New York: Doubleday.

Normile, D. (2000). "Hopes Grow for Hybrid Rice to Feed Developing World." *Science* 288: 429.

Orwell, G. (1982). *1984.* New York: Buccaneer Books.

Orwell, G. (1996). *Animal Farm: A Fairy Story.* 50th anniversary ed. New York: Harcourt Brace.

Pennisi, E. (1999). "Do Mitochondrial Mutations Dim the Fire of Life?" *Science* 286: 664.

Pennisi, E. (2000a). "Mouse Sequencers Take Up the Shotgun." *Science* 287: 1179–1181.

Pennisi, E. (2000b). "Fruit Fly Genome Yields Data and a Validation." *Science* 287: 1374.

Pennisi, E. (2000c). "Ideas Fly at Gene-Finding Jamboree." *Science* 287: 2182–2184.

Pennisi, E. (2000d). "Stealth Genome Rocks Rice Researchers." *Science* 288: 239–241.

Pennisi, E. (2000e). "DOE Team Sequences Three Chromosomes." *Science* 288: 417–419.

Pennisi, E. (2000f). "Chromosome 21 Done, Phase Two Begun." *Science* 288: 939.

Pennisi, E. (2000g). "And the Gene Number Is . . . ?" *Science* 288: 1146–1147.

Pennisi, E. (2000h). "Finally, the Book of Life and Instructions for Navigating It." *Science* 288: 2304–2307.

Pennisi, E. (2000i). "Rat Genome Off to an Early Start." *Science* 289: 1267–1269.

Pennisi, E., and Vogel, G. (2000). "Clones: A Hard Act to Follow." *Science* 288: 1722–1727.

Pennisi, E., and Normile, D. (2000). "Perseverance Leads to Cloned Pigs in Japan." *Science* 289: 118–119.

Perry, D. (2000). "Patients' Voices: The Powerful Sound in the Stem Cell Debate." *Science* 287: 1423.

Reich, M. R. (2000). "The Global Drug Gap." *Science* 287: 1979–1981.

Roberts, L. (2000). "SNP Mappers Confront Reality and Find It Daunting." *Science* 287: 1898–1899.

Rosamond, J., and Allsop, A. (2000). "Harnessing the Power of the Genome in the Search for New Antibiotics." *Science* 287: 1973–1976.

Rossner, J. (1977). *Attachments.* New York: Simon & Schuster.

Ruestow, E. G. (1996). *The Microscope in the Dutch Republic—The Shaping of Discovery.* Cambridge, UK: Cambridge University Press.

Ryder, O. A., McLaren, A., Brenner, S., Zhang, Y. P., and Benirschke, K. (2000). "DNA Banks for Endangered Animal Species." *Science* 288: 275–277.

Sacks, O. (1987). *The Man Who Mistook His Wife for a Hat and Other Clinical Tales.* New York: Harper & Row.

Sander, C. (2000). "Genomic Medicine and the Future of Health Care." *Science* 287: 1977–1978.

Scott, E. C. (2000). "Not (Just) in Kansas Anymore." *Science* 288: 813–815.

Seife, C. (2000). "'Ultimate PC' Would Be A Hot Little Number." *Science* 289: 1447–1448.

Service, R. F. (1998). "Microchip Arrays Put DNA on the Spot." *Science* 282: 396–399.

Service, R. F. (1999a). "Neurons and Silicon Get Intimate." *Science* 284: 578–579.

Service, R. F. (1999b). "Big Blue Aims to Crack Protein Riddle." *Science* 286: 2250.

Service, R. F. (1999c). "Building the Small World of the Future." *Science* 286: 2442–2444.

Service, R. F. (2000a). "Nanocrystals May Give Boost to Data Storage." *Science* 287: 1902–1903.

Service, R. F. (2000b). "Structural Genomics Offers High-Speed Look at Proteins." *Science* 287: 1954–1956.

Service, R. F. (2000c). "Can Celera Do It Again?" *Science* 287: 2136–2138.

Service, R. F. (2000d). "Creation's Seventh Day." *Science* 289: 232–235.

Service, R. F. (2000e). "Tissue Engineers Build New Bone." *Science* 289: 1498–1500.

Service, R. F. (2000f). "Protein Arrays Step Out of DNA's Shadow." *Science* 289: 1673.

Sikorski, R., and Peters, R. (1999). "Test Tube Kidneys." *Science* 284: 1952.

Slack, J. M. W. (2000). "Stem Cells in Epithelial Tissues." *Science* 287: 1431–1433.

Smaglik, P. (2000). "Researchers Take a Gamble on the Human Genome." *Nature* 405: 264.

Solter, D., and Gearhart, J. (1999). "Putting Stem Cells to Work." *Science* 283: 1468–1470.

Steghaus-Kovac, S. (1999a). "Stem Cells as Potential Nerve Therapy." *Science* 285: 650.

Steghaus-Kovac, S. (1999b). "Ethical Loophole Closing Up for Stem Cell Researchers." *Science* 286: 31.

Stone, R. (2000). "U.K. Backs Use of Embryos, Sets Vote." *Science* 289: 1269–1270.

Strauss, E. (1999a). "Can Mitochondrial Clocks Keep Time"? *Science* 283: 1435–1438.

Strauss, E. (1999b). "mtDNA Shows Signs of Paternal Influence." *Science* 286: 2436.

Strauss, E. (2000). "New Clue to Age Control in Yeast." *Science* 287: 1181–1182.

Taubes, G. (2000). "Biologists and Engineers Create a New Generation of Robots That Imitate Life." *Science* 288: 80–83.

Twain, M. (1999). *Pudd'Nhead Wilson and Other Tales: Those Extraordinary Twins, the Man That Corrupted Hadleyburg*. New York: Oxford University Press.

Van der Kooy, D., and Weiss, S. (2000). "Why Stem Cells?" *Science* 287: 1439–1441.

Venter, J. C. et al. (2001). "The Sequence of the Human Genome." *Science* 291: 1304–1351.

Vian, B. (1948). *Et On Tuera Tous les Affreux*. Paris: Pauvert.

Vincent, W. F. (1999). "Icy Life on a Hidden Lake." *Science* 286: 2094–2095.

Vogel, G. (1999a). "NIH Sets Rules for Funding Embryonic Stem Cell Research." *Science* 286: 2050–2051.

Vogel, G. (1999b). "Capturing the Promise of Youth." *Science* 286: 2238–2239.

Vogel, G. (1999c). "Mice Cloned from Cultured Stem Cells." *Science* 286: 2437.

Vogel, G. (2000a). "Company Gets Rights to Cloned Human Embryos." *Science* 287: 559.

Vogel, G. (2000b). "Wisconsin to Distribute Embryonic Stem Cells." *Science* 287: 948–949.

Vogel, G. (2000c). "Can Old Cells Learn New Tricks?" *Science* 287: 1418–1419.

Vogel, G. (2000d). "Congress Investigates Fetal Tissue Sales." *Science* 287: 1903–1905.

Vogel, G. (2000e). "Tracking the Movements That Shape an Embryo." *Science* 288: 86–87.

Vogel, G. (2000f). "In Contrast to Dolly, Cloning Resets Telomere Clock in Cattle." *Science* 288: 586–587.

Vogel, G. (2000g). "Brain Cells Reveal Surprising Versatility." *Science* 288: 1559–1561.

Vogel, G. (2000h). "Researchers Get Green Light for Work on Stem Cells." *Science* 289: 1442–1443.

Vogel, G (2000i). "Brain Cells Turning Over a New Leaf." *Science* 289: 1666.

Waterson, R., and Sulston, J. E. (1998). "The Human Genome Project: Reaching the Finish Line." *Science* 282: 53–54.

Watson, J. D., Hopkins, N., Roberts, J., Argetsinger Steitz, J., and Weiner, A. (1988). *Molecular Biology of the Gene*. 4th ed. Menlo Park, CA: Benjamin/Cummings.

Watt, F. M., and Hogan, B. L. M. (2000). "Out of Eden: Stem Cells and Their Niches." *Science* 287: 1427–1430.

Weissman, I. L. (2000). "Translating Stem and Progenitor Cell Biology to the Clinic: Barriers and Opportunities." *Science* 287: 1442–1446.

Wickelgren, I. (1999). "Mining the Genome for Drugs." *Science* 285: 998–1001.

Wiener, N. (1948). *Cybernetics, or Control and Communication in the Animal and the Machine*. New York: John Wiley.

Young, F. E. (2000). "A Time for Restraint." *Science* 287: 1424.

Zanjani, E. D., and French Anderson, W. (1999). "Prospects for in Utero Human Gene Therapy." *Science* 285: 2084–2087.

## Technical References

Adami, C., Belew, R. K., Kitano, H., and Taylor, C. E. (1998). *Artificial Life VI: Proceedings of the Sixth International Conference on Artificial Life*. Cambridge, Mass.: MIT Press.

Adams, M. D., Celniker, S. E., Holt, R. A., et al. (2000). "The Genome Sequence of *Drosophila melanogaster*." *Science* 287: 2185–2195.

Apfeld, J., and Kenyon, C. (1999). "Regulation of Lifespan by Sensory Perception in *Caenorhabditis elegans*." *Nature* 402: 804–809.

Ariieumi, T., Komazaki, S., Asashima, M., and Malacinski, G. M. (1996). "Activin Treated Urodele Ectoderm: A Model Experimental System for Cardiogenesis." *Journal of Developmental Biology* 40: 715–718.

Baguisi, A., Behboodi, E., Melican, D. T., Pollock, J. S., Destrempes, M. M., Cammuso, C., Williams, J. L., Nims, S. D., Porter, C. A., Midura, P., Palacios, M. J., Ayres, S. L., Denniston, R. S., Hayes, M. L., Ziomek, C. A., Meade, H. M., Godke, R. A., Gavin, W. G., Overstrom, E. W., and Echelard, Y. (1999). "Production of Goats by Somatic Cell Nuclear Transfer." *Nature Biotechnology* 17: 456–461.

Baldi, P. and Brunak, S. (1998). *Bioinformatics: The Machine Learning Approach*. Cambridge, Mass.: MIT Press. Second edition (2001).

Baldi, P., Brunak, S., Frasconi, P., Soda, G., and Pollastri, G. (1999). "Exploiting the Past and the Future in Protein Secondary Structure Prediction." *Bioinformatics* 15: 937–946.

Blelloch, R., and Kimble, J. (1999). "Control of Organ Shape by a Secreted Metalloprotease in the Nematode *Caenorhabditis elegans*." *Nature* 399: 586–590.

Briggs, R., and King, T. J. (1952). "Transplantation of Living Cell Nuclei from Blastula Cells into Enucleated Frog's Eggs." *Proceedings of the National Academy of Sciences USA* 38: 455–463.

Brüstle, O., Jones, K. N., Learish, R. D., Karram, K., Choudhary, K., Wiestler, O. D., Duncan, I. D., and McKay, R. D. G. (1999). "Embryonic Stem Cell-Derived Glial Precursors: A Source of Myelinating Transplants." *Science* 285: 754–756.

Bush, R. M., Bender, C. A., Subbarao, K., Cox, N. J., and Fitch, W. M. (1999). "Predicting the Evolution of Human Influenza A." *Science* 286: 1921–1924.

The *C. elegans* Sequencing Consortium (1998). "Genome Sequence of the Nematode *C. elegans:* A Platform for Investigating Biology." *Science* 282: 2012–2018.

Carrier, R. L., Papadaki, M., Rupnick, M., Schoen, F. J., Bursac, N., Langer, R., Freed, L. E., and Vunjak-Novakovic, G. (1999). "Cardiac Tissue Engineering: Cell Seeding, Cultivation Parameters, and Tissue Construct Characterization." *Biotechnology and Bioengineering* 64: 580–589.

Chan, A. W. S., Dominko, T., Luetjens, C. M., Neuber, E., Martinovich, C., Hewitson, L., Simerly, C. R., and Schatten, G. P. (2000). "Clonal Propagation of Primate Offspring by Embryo Splitting." *Science* 287: 317–319.

Clarke, D. L., Johansson, C. B., Wilbertz, J., Veress, B., Nilsson, E., Karlström, H., Lendahl, U., and Frisén, J. (2000). "Generalized Potential of Adult Neural Stem Cells." *Science* 288: 1660–1663.

Eriksson, P. S., Perfilieva, E., Bjork-Eriksson, T., Alborn, A. M., Nordborg, C., Peterson, D. A., and Gage, F. H. (1998). "Neurogenesis in the Adult Human Hippocampus." *Nature Medicine* 4: 1313–1317.

Evans, M. J., Gurer, C., Loike, J. D., Wilmut, I., Schnieke, A. E., and Schon, E. A. (1999). "Mitochondrial DNA Genotypes in Nuclear Transfer-Derived Cloned Sheep." *Nature Genetics* 23: 90–93.

Fisher, P. E., Russell, D. A., Stoskopf, M. K., Barrick, R. E., Hammer, M., and Kuzmitz, A. A. (2000). "Cardiovascular Evidence for an Intermediate or Higher Metabolic Rate in an Ornithischian Dinosaur." *Science* 288: 503–505.

Francis, D., Diorio, J., Liu, D., and Meaney, M. J. (1999). "Nongenomic Transmission Across Generations of Maternal Behavior and Stress Response in the Rat." *Science* 286: 1155–1158.

Frary, A., Clint Nesbitt, T., Frary, A., Grandillo, S., van der Knaap, E., Cong, B., Liu, J., Meller, J., Elber, R., Alpert, K. B., and Tanksley, S. D. (2000). "*fw2.2:* A Quantitative Trait Locus Key to the Evolution of Tomato Fruit Size." *Science* 289: 85–88.

Freed, W. J. (2000). *Neural Transplantation: An Introduction*. Cambridge: MIT Press.

Garcia, J. A., Zhang, D., Estill, S. J., Michnoff, C., Rutter, J., Reick, M., Scott, K., Diaz-Arrastia, R., and McKnight, S. L. (2000). "Impaired Cued and Contextual Memory in NPAS2-Deficient Mice." *Science* 288: 2226–2230.

Garey, M. R., and Johnson, D. S. (1979). *Computers and Intractability*. New York: W. H. Freeman.

Gould, E., Reeves, A. J., Graziano, M. S., and Gross, C. G. (1999). "Neurogenesis in the Neocortex of Adult Primates." *Science* 286: 548–552.

Griffith, M., Osborne, R., Munger, R., Xiong, X., Doillon, C. J., Laycock, N. L. C., Hakim, M., Song, Y., and Watsky, M. A. (1999). "Functional Human Corneal Equivalents Constructed from Cell Lines." *Science* 286: 2169–2172.

Gurdon, J. B., (1962). "Adult Frogs Derived from the Nuclei of Somatic Cells." *Developmental Biology* 4: 256–273.

Gurdon, J. B. (1986). "Nuclear Transplantation in Eggs and Oocytes." *Journal of Cell Science* (suppl.) 4: 287–318.

Gurdon, J. B., and Uehlinger, V. (1966). "'Fertile' Intestine Nuclei." *Nature* 210: 1240–1241.

Hattori, M., Fujiyama, A., Taylor, T. D., et al. (2000). "The DNA Sequence of Human Chromosome 21." *Nature* 405: 311–319.

Herzog, R. W., Yang, E. Y., Couto, L. B., Hagstrom, J. N., Elwell, D., Fields, P. A., Burton, M., Bellinger, D. A., Read, M. S., Brinkhous, K. M., Podsakoff, G. M., Nichols, T. C., Kurtzman, G. J., and High, K. A. (1999). "Long-Term Correction of Canine Hemophilia B by Gene Transfer of Blood Coagulation Factor IX Mediated by Adeno-Associated Viral Vector." *Nature Medicine* 5: 56–63.

Hutchinson, C. A. III, Peterson, S. N., Gill, S. R., Cline, R. T., White, O., Fraser, C. M., Smith, H. O., and Venter, J. C. (1999). "Global Transposon Mutagenesis and a Minimal Mycoplasma Genome." *Science* 286: 2165–2169.

Illmensee, K. (1968). "Transplantation of Embryonic Nuclei into Unfertilized Eggs of *Drosophila melanogaster*." *Nature* 219: 1268–1269.

Illmensee, K., and Hoppe, P. C. (1981). "Nuclear Transplantation in *Mus musculus*: Developmental Potential of Nuclei from Preimplantation Embryos." *Cell* 23: 9–18.

Illmensee, K., Hoppe, P. C., and Croce, C. M. (1978). "Chimeric Mice Derived from Human-Mouse Hybrid Cells." *Proceedings of the National Academy of Sciences USA* 75: 1914–1918.

Jung, J., Zheng, M., Goldfarb, M., and Zaret, K. S. (1999). "Initiation of Mammalian Liver Development from Endoderm by Fibroblast Growth Factors." *Science* 284: 1998–2003.

Kaesseman, H., Wiebe, V., and Pääbo, S. (1999). "Extensive Nuclear DNA Sequence Diversity Among Chimpanzees." *Science* 286: 1159–1162.

Kato, Y., Tani, T., Sotomaru, Y., Kurokawa, K., Kato, J., Doguchi, H., Yasue, H., and Tsunoda, Y. (1998). "Eight Calves Cloned from Somatic Cells of a Single Adult." *Science* 282: 2095–2098.

Kay, M. A., Manno, C. S., Ragni, M. V., Larson, P. J., Couto, L. B., McClelland, A., Glader, B., Chew, A. J., Tai, S. J., Herzog, R. W., Arruda, V., Johnson, F., Scallan, C., Skarsgard, E., Flake, A. W., and High, K. A. (2000). "Evidence for

Gene Transfer and Expression of Factor IX in Haemophilia B Patients Treated with an AAV Vector." *Nature Genetics* 24: 257–261.

Kornberg, T. B., and Krasnow, M. A. (2000). "The *Drosophila* Genome Sequence: Implications for Biology and Medicine." *Science* 287: 2218–2220.

Kotz, R., Windhager, R., Dominkus, M., Robioneck, B., and Müller-Daniels, H. (2000). "A Self-Extending Paediatric Leg Implant." *Nature* 406: 143–144.

Krings, M., Geisert, H., Schmitz, R. W., Krainitzki, H., and Pääbo, S. (1999). "DNA Sequence of the Mitochondrial Hypervariable Region II from the Neanderthal Type Specimen." *Proceedings of the National Academy of Sciences USA* 96: 5581–5585.

Krings, M., Stone, A., Schmitz, R. W., Krainitzki, H., Stoneking, M., and Pääbo, S. (1997). "Neanderthal DNA Sequences and the Origin of Modern Humans." *Cell* 90: 19–30.

Langton, C. (1987). *Artificial Life*. Wakefield, Mass.: Addison-Wesley.

Lanza, R. P., Cibelli, J. B., Blackwell, C., Cristofalo, V. J., Francis, M. K., Baerlocher, G. M., Mak, J., Schertzer, M., Chavez, E. A., Sawyer, N., Lansdorp, P. M., and West, M. D. (2000). "Extension of Cell Life-Span and Telomere Length in Animals Cloned from Senescent Somatic Cells." *Science* 288: 665–669.

Lee, C., Klopp, R. G., Weindruch, R., and Prolla, T. A. (1999). "Gene Expression Profile of Aging and Its Retardation by Caloric Restriction." *Science* 285: 1390–1393.

Lin, S. J., Defossez, P. A., and Guarente, L. (2000). "Requirement of NAD and *SIR2* for Life-Span Extension by Calorie Restriction in *Saccharomyces cerivisiae*." *Science* 289: 2126–2128.

Lin, Y. J., Seroude, L., and Benzer, S. (1998). "Extended Life-Span and Stress Resistance in the *Drosophila* Mutant Methuselah." *Science* 282: 943–946.

Lloyd, S. (2000). "Ultimate Physical Limits to Computation." *Nature* 406: 1047–1054.

Lowenstein, D. H., and Parent, J. M. (1999). "Brain, Heal Thyself." *Science* 283: 1126–1127.

Magavi, S. S., Leavitt, B. R., and Macklis, J. D. (2000). "Induction of Neurogenesis in the Neocortex of Adult Mice." *Nature* 405: 951–955.

Maher, M. P., Pine, J., Wright, J., and Tai, Y. C. (1999). "The Neurochip: A New Multielectrode Device for Stimulating and Recording from Cultured Neurons." *Journal of Neuroscience Methods* 87: 45–56.

Maletic-Savatic, M., Malinow, R., and Svoboda, K. (1999). "Rapid Dendritic Morphogenesis in CA1 Hippocampal Dendrites Induced by Synaptic Activity." *Science* 283: 1923–1927.

McDonald, J. W., Liu, X. Z., Qu, Y., Liu, S., Mickey, S. K., Turetsky, D., Gottlieb, D. I., and Choi, D. W. (1999). "Transplanted Embryonic Stem Cells Survive,

Differentiate and Promote Recovery in Injured Rat Spinal Cord." *Nature Medicine* 5: 1410–1412.

McGrath, J., and Solter, D. (1983). "Nuclear Transplantation in the Mouse Embryo by Microsurgery and Cell Fusion." *Science* 220: 1300–1302.

Mead, C. (1989). *Analog VLSI and Neural Systems.* Wakefield, Mass.: Addison-Wesley.

Michikawa, Y., Mazzucchelli, F., Bresolin, N., Scarlato, G., and Attardi, G. (1999). "Aging-Dependent Large Accumulation of Point Mutations in the Human mtDNA Control Region for Replication." *Science* 286: 774–779.

Niklason, L. E., Gao, J., Abbott, W. M., Hirschi, K. K., Houser, S., Marini, R., and Langer, R. (1999). "Functional Arteries Grown in Vitro." *Science* 284: 489–493.

Ogawa, T., Dobrinski, I., Avarbock, M. R., and Brinster, R. L. (2000). "Transplantation of Male Germ Line Stem Cells Restores Fertility in Infertile Mice." *Nature Medicine* 6: 29–34.

Okabe, M., Ikawa, M., Kominami, K., Nakanishi, T., and Nishimune, Y. (1997). "'Green Mice' as a Source of Ubiquitous Green Cells." *FEBS Letters* 407: 313–319.

Onishi, A., Iwamoto, M., Akita, T., Mikawa, S., Takeda, K., Awata, T., Hanada, H., and Perry, A. C. F. (2000). "Pig Cloning by Microinjection of Fetal Fibroblast Nuclei." *Science* 289: 1188–1190.

Ovchinnikov, I. V., Götherström, A., Romanova, G. P., Kharitonov, V. M., Lidén, K., and Goodwin, W. (2000). "Molecular Analysis of Neanderthal DNA from the Northern Caucasus." *Nature* 404: 490–493.

Palermo, G., Joris, H., Devroey, P., and Van Steirteghem, A. C. (1992). "Pregnancies after Intracytoplasmic Injection of Single Spermatozoon into an Oocyte." *Lancet* 340: 17–18.

Paradis, K., Langford, G., Long, Z., Heneine, W., Sandstrom, P., Switzer, W. M., Chapman, L. E., Lockey, C., Onions, D., the XEN 111 Study Group, and Otto, E. (1999). "Search for Cross-Species Transmission of Porcine Endogenous Retrovirus in Patients Treated with Living Pig Tissue." *Science* 285: 1236–1241.

Perry, A. C. F., Wakayama, T., Kishikawa, H., Kasai, T., Okabe, M., Toyoda, Y., and Yanagimachi, R. (1999). "Mammalian Transgenesis by Intracytoplasmic Sperm Injection." *Science* 284: 1180–1183.

Petersen, B. E., Bowen, W. C., Patrene, K. D., Mars, W. M., Sullivan, A. K., Murase, N., Boggs, S. S., Greenberger, J. S., and Goff, J. P. (1999). "Bone Marrow as a Potential Source of Hepatic Oval Cells." *Science* 284: 1168–1170.

Phillips, R. L., Ernst, R. E., Brunk, B., Ivanova, N., Mahan, M. A., Deanehan, J. K., Moore, K. A., Overton, G. C., and Lemischka, I. R. (2000). "The Genetic Program of Hematopoietic Stem Cells." *Science* 288: 1635–1640.

Prather, R. S., Barnes, F. L., Sims, M. L., Robl, J. M., Eyestone, W. H., and First, N. L. (1987). "Nuclear Transfer in the Bovine Embryo: Assessment of Donor Nuclei and Recipient Oocyte." *Biology of Reproduction* 37: 859–866.

Prather, R. S., Sims, M. L., and First, N. L. (1989). "Nuclear Transplantation in Early Pig Embryos." *Biology of Reproduction* 41: 414–418.

Pratt, J. W., Raiffa, H., and Schlaifer, R. (1996). *Statistical Decision Theory*. Cambridge, Mass.: MIT Press.

Ramiya, V. K., Maraist, M., Arfors, K. E., Schatz, D. A., Peck, A. B., and Cornelius, J. G. (2000). "Reversal of Insulin-Dependent Diabetes Using Islets Generated in Vitro from Pancreatic Stem Cells." *Nature Medicine* 6: 278–282.

Rampon, C., Tang, Y. P., Goodhouse, J., Shimizu, E., Kyin, M., and Tsien, J. Z. (2000). "Enrichment Induces Structural Changes and Recovery from Nonspatial Memory Deficits in CA1 NMDAR1-Knockout Mice." *Nature Neuroscience* 3: 238–244.

Rong, Y. S., and Golic, K. G. (2000). "Gene Targeting by Homologous Recombination in *Drosophila*." *Science* 288: 2013–2018.

Rubin, G. M., Yandell, M. D., Wortman, J. R., et al. (2000). "Comparative Genomics of the Eukaryotes." *Science* 287: 2204–2215.

Schnieke, A. E., Kind, A. J., Ritchie, W. A., Mycock, K., Scott, A. R., Ritchie, M., Wilmut, I., Colman, A., and Campbell, K. H. S. (1997). "Human Factor IX Transgenic Sheep Produced by Transfer of Nuclei from Transfected Fetal Fibroblasts." *Science* 278: 2130–2133.

Schulz, W. W., Van Andel, P., Sabelis, I., and Mooyaart, E. (1999). "Magnetic Resonance Imaging of Male and Female Genitals During Coitus and Female Sexual Arousal." *British Medical Journal* 319: 1596–1600.

Shearman, L. P., Sriram, S., Weaver, D. R., et al. (2000). "Interacting Molecular Loops in the Mammalian Circadian Clock." *Science* 288: 1013–1019.

Shi, S. H., Hayashi, Y., Petralia, R. S., Zaman, S. H., Wenthold, R. J., Svoboda, K., and Malinow, R. (1999). "Rapid Spine Delivery and Redistribution of AMPA Receptors after Synaptic NMDA Receptor Activation." *Science* 284: 1811–1816.

Shiels, P. G., Kind, A. J., Campbell, K. S., Waddington, D., Wilmut, I., Colman, A., and Schnieke, A. E. (1999). "Analysis of Telomere Lengths in Cloned Sheep." *Nature* 399: 316.

Shihabuddin, L. S., Brunschwig, J. P., Holets, V. R., Bunge, M. B., and Whittemore, S. R. (1996). Induction of Mature Neuronal Properties in Immortalized Neuronal Precursor Cells Following Grafting into the Neonatal CNS. *Journal of Neurocytology* 25: 101–111

Smith, L. D. (1965). "Transplantation of the Nuclei of Primordial Germ-Cells into Enucleated Eggs of *Rana pipiens*." *Proceedings of the National Academy of Sciences USA* 54: 101–107.

Tang, Y., Shimizu, E., Dube, G. R., Rampon, C., Kerchner, G. A., Zhuo, M., Liu, G., and Tsien, J. Z. "Genetic Enhancement of Learning and Memory in Mice." *Nature* 401: 63–69.

Thomson, J. A., Itskovitz-Eldor, J., Shapiro, S. S., Waknitz, M. A., Swiergiel, J. J., Marshall, V. S., and Jones, J. M. (1998). "Embryonic Stem Cell Lines Derived from Human Blastocysts." *Science* 282: 1145–1147.

Tomizuka, K., Yoshida, H., Uejima, H., Kugoh, H., Sato, K., Ohguma, A., Hayasaka, M., Hanaoka, K., Oshimura, M., and Ishida, I. (1997). "Functional Expression and Germline Transmission of a Human Chromosome Fragment in Chimaeric Mice." *Nature Genetics* 16: 133–143.

Tropepe, V., Coles, B. L. K., Chiasson, B. J., Horsford, D. J., Elias, A. J., McInnes, R. R., and van der Kooy, D. (2000). "Retinal Stem Cells in the Adult Mammalian Eye." *Science* 287: 2032–2036.

Wakayama, T., Perry, A. C., Zuccotti, M., Johnson, K. R., and Yanagimachi, R. (1998). "Full-Term Development of Mice from Enucleated Oocytes Injected with Cumulus Cell Nuclei." *Nature* 394: 369–374.

Wakayama, T. W., and Yanagimachi, R. (1999). "Cloning of Male Mice from Adult Tail-Tip Cells." *Nature Genetics* 22: 127–128.

Walker, D. W., McColl, G., Jenkins, N. L., Harris, J., and Lithgow, G. J. (2000). "Natural Selection: Evolution of Lifespan in *C. elegans*." *Nature* 405: 296–297.

Watson, J. D., and Crick, F. H. C. (1953). "Molecular Structure of Nucleic Acids." *Nature* 171: 737.

White, O., Eisen, J. A., Heidelberg, J. F., et al. (1999). "Genome Sequence of the Radioresistant Bacterium *Deinococcus radiodurans* R1." *Science* 286: 1571–1577.

Willadsen, S. M. (1979). "A Method for Culture of Micromanipulated Sheep Embryos and Its Use to Produce Monozygotic Twins." *Nature* 277: 298–300.

Willadsen, S. M. (1986). "Nuclear Transplantation of Sheep Embryos." *Nature* 320: 63–65.

Wilmut, I., Campbell, K. H. S., Schniekel, A. E., McWhir, J., and Kind, A. J. (1997). "Viable Offspring Derived from Fetal and Adult Mammalian Cells." *Nature* 385: 810–813.

Woodbury, D., Schwarz, E. J., Prockop, D. J., and Black, I. B. (2000). "Adult Rat and Human Bone Marrow Stromal Cells Differentiate into Neurons." *Journal of Neuroscience Research* 61: 364–370.

Yagi, M., Ritchie, K. A., Sitnicka, E., Storey, C., Roth, G. J., and Bartelmez, S. (1999). "Sustained ex Vivo Expansion of Hematopoietic Stem Cells Mediated by Thrombopoietin." *Proceedings of the National Academy of Sciences USA* 96: 8126–8131.

Yandava, D., Billinghurst, L. L., and Snyder, E. Y. (1999). "'Global' Cell Replacement Is Feasible via Neural Stem Cell Transplantation: Evidence from the Dysmyelinated Shiverer Mouse Brain." *Proceedings of the National Academy of Sciences USA* 96: 7029–7034.

Ye, X., Al-Babili, S., Klöti, A., Zhang, J., Lucca, P., Beyer, P., and Potrykus, I. (2000). "Engineering the Provitamin A (β-Carotene) Biosynthetic Pathway into (Carotenoid-Free) Rice Endosperm." *Science* 287: 303–305.

Zamanillo, D., Sprengel, R., Hvalby, Ø., Jensen, V., Burnashev, N., Rozov, A., Kaiser, K. M. M. , Köster, H. J., Borchardt, T., Worley, P., Lübke, J., Frotscher, M., Kelly, P. H., Sommer, B., Andersen, P., Seeburg, P. H., and Sakmann, B. (1999). "Importance of AMPA Receptors for Hippocampal Synaptic Plasticity but Not for Spatial Learning." *Science* 284: 1805–1811.

Zimmerman, U., and Vienken, J. (1982). "Electric-Field-Induced Cell-to-Cell Fusion." *Journal of Membrane Biology* 67: 165–182.

# Index